T0074593

The Science Matrix

Frederick Seitz

The Science Matrix

The Journey, Travails, Triumphs

Springer-Verlag

New York Berlin Heidelberg London Paris
Tokyo Hong Kong Barcelona Budapest

Frederick Seitz
The Rockefeller University
New York, NY 10021-6399
USA

Library of Congress Cataloging-in-Publication Data
Seitz, Frederick, 1911–
    The science matrix:the journey, travails, triumphs/Frederick
Seitz.
        p.   cm.
    Includes bibliographical references and index.
    ISBN-13: 978-0-387-98574-9   e-ISBN-13: 978-1-4612-2828-8
    DOI: 10.1007/978-1-4612-2828-8

    1. Science—Popular works.   2. Technology—Popular works.
3. Science—History—Popular works.   4. Technology—History—Popular
works.   I. Title.
Q162.S454   1992
500—dc20   91-27828

Printed on acid-free paper.

Production managed by Natalie Johnson; manufacturing supervised by Jacqui Ashri.
Photocomposed copy prepared from the author's WordPerfect files using Ventura Publisher.

9 8 7 6 5 4 3 2 1

To E.M.S.
Lifetime Companion and Valued Critic

Many the wonders but nothing walks stranger than man.
This thing crosses the sea in the winter's storm,
making his path through the roaring waves.
And she, the greatest of goods, the earth—
ageless she is, and unwearied—he wears her away
as the ploughs go up and down from year to year
and his mules turn up the soil.

Gay nations of birds he snares and leads,
wild beast tribes and the salty brood of the sea,
with the twisted mesh of his nets, this clever man.
He controls with craft the beasts of the open air,
walkers on hills. The horse with his shaggy mane
he holds and harnesses, yoked about the neck,
and the strong bull of the mountain.

Language, and thought like the wind
and the feelings that make the town,
he has taught himself, and shelter against the cold,
refuge from rain. He can always help himself.
He faces no future helpless. There's only death
that he cannot find an escape from. He has contrived
refuge from illnesses once beyond all cure.

Clever beyond all dreams
the inventive craft that he has
which may drive him one time or another to well or ill.
When he honors the laws of the land and the gods' sworn right
high indeed is his city; but stateless the man
who dares to dwell with dishonor. Not by my fire,
never to share my thoughts, who does these things.

Greek chorus in *Antigone* by Sophocles
Translation by Richard Lattimore and David Grene
Complete Greek Tragedies
(University of Chicago Press, 1957)

Preface

The primary object of this small book is to emphasize in a somewhat personal way how the natural sciences arose and what they mean to our present civilization—with all their positives and negatives. Someone asked if a primary goal of the book is also to press for more money for the support of science. This is definitely not the case! The advance of science requires money given with appreciation and wisdom but the amounts must be determined by many complex factors which lie outside the scope of this book. I hope, however, that the essays will make it clear to those who read them that the future well-being of humanity will inevitably be closely linked to the advance of both science and science-based technology. We have gone too far in depending upon these twins born out of Renaissance culture to turn back. To do so would guarantee that all mankind would lapse into the dismal fifth stage of civilization which Charles G. Darwin feared to be our inevitable fate, as is described in Chapter 7.

In a sense the pivotal chapter of the book which reflects a lifelong interest in the history of science is Chapter 3. It deals with the way in which science reached maturity—a wonder and yet not a wonder, to paraphrase the Dutch physicist-engineer, Simon Stevin, when he discovered the law of equilibrium of forces. However, the other chapters play their own supporting roles.

Chapter 3, which I have termed a central one, is not a formal study in the history of science but I feel confident that many individuals, including most fellow scientists, will find it worth reading to obtain a better understanding of how we reached the point we are at present.

There is a school of thought today pervading segments of many universities, particularly some of the most prominent ones in the United States, that far too much attention has been given to the influence of the Greeks of the ages of Pythagoras, Plato, and Aristotle and that their work was merely a reflection of the perhaps equal or even greater accomplishments of individuals in older civilizations. Far be it from me to demean the great achievements of the civilizations that arose in upper and lower Egypt, Mesopotamia, Persia, India, and China, as well as those of the Mayans and Incas. However, anyone at all familiar with the type of inspired insight that a great step in the advance of science requires can fully appreciate the entirely unique flame that was set alight by the Athenians and those immediately influenced by them. All this goes with the understanding that in this period Athens was a melting pot both ethnically and culturally.

Essentially all civilizations that rose to the level of possessing an urban culture had need for two forms of science-related technology, namely, mathematics for land measurements and commerce and astronomy for time-keeping in agriculture and aspects of religious rituals. Exactly how far in sophistication those fields were pursued in each civilization depended upon a local combination of interest, leisure, and wealth. Moreover, there is inevitably a certain degree of commonality because most civilizations are partially derivative through diffusion and partially original.

What the Greeks did in determining on experimental grounds that the earth is a sphere, in evaluating its radius, in estimating the distance to the moon, and in speculating about the nature of the solar system, however, was unique for its time and very different from what had gone before. They in effect turned the system on its head in searching for a deeper understanding of the natural world. It was, in fact, this collective work which, when transferred to Europe with augmentation through the Arabs, provided the inspiration for individuals such as Buridan, Copernicus, Galileo, Kepler, and those who followed them during the European Renaissance. Moreover, the Greek contribution to world culture was of a general and not parochial nature. They proffered gifts of a universal kind which, in the long run, have substantially influenced all civilizations. To deny all this, as is sometimes done these days, is simply bad scholarship and will stand forth as such in the last analysis.

Several of the essays have appeared in somewhat different form in special journals and one appeared in a much earlier version in a multi-authored book of articles. I am grateful to the several publishers for permitting me to adapt them for the purposes of this book.

As might be expected, I am indebted to many friends and other scholars for much help. I am particularly indebted to Drs. A. G. Bearn, E. R. Piore, and Professor Miro Todorovich for critical comments. I am also indebted to The Reverend Armand A. Maurer, CSB, for authoritative information concerning St. Augustine's views on Aristotle's mechanics. Detailed statements concerning help received from Professors S. Chandrasekhar and A.I. Sabra appear in the leading footnote to Chapter 3. Not least I am indebted to Mrs. Florence Arwade for guiding the entire program through its many phases and to Mrs. Sonya Mirsky for innumerable fruitful discussions of aspects of the history of science.

Finally, I would like to give very special thanks to the editorial department of Springer-Verlag for its critical reading of the entire book.

Acknowledgment of Sources

Several of the essays have appeared in modified form in other pieces.

Chapter 2 is a revision of an essay by the author which appeared in *Science and Technology in the World of the Future* edited by Arthur B. Bronwell (Wiley Interscience, New York, 1970).

Chapters 3 and 8 appeared in slightly different form in the *International Journal on the Unity of the Sciences*, Volume 24, number 1 (1991) and Volume 2, number 3 (1989) respectively.

The quotation from Sophocles' *Antigone* in the frontispiece is in accordance with the translations of Richard Lattimore and David Grene, *Complete Greek Tragedies* (University of Chicago Press, 1954, 1957).

The translation of the statement by Ibn Khaldun at the beginning of Section 8 of Chapter 3 was made by Professor Bernard Lewis of Princeton University whose book is listed in the bibliography going with that chapter.

As noted, the quotation from Immanuel Kant's *Essay on Eternal Peace* which appears at the heading of Section 14 of Chapter 7 is from the translation by Lewis White Beck.

Contents

1
The Brain Matrix:
Our Window on the World

"Truth is a remarkable thing. We cannot miss knowing some of it. But we cannot know it entirely." Aristotle

(Free translation of inscription on the building of the National Academy of Sciences)

The mathematicians have, among many other things, an invention termed the transformation matrix. This can be used in numerous ways. For example, it can be used to describe aspects of a change in the position of an object or it can provide a link between two rather different systems of things. Consider a very simple case of the former kind. You are sitting at your desk in a swivel chair and rotate through 90o in order to get up. There is a transformation matrix that can describe the rotation. Or, to give an example of the second kind, if you are planning to map the surface of a spherical globe describing the surface features of the earth onto a flat sheet there are transformation matrices which can provide several useful and interesting ways to do it. The matrix can also play much more complicated mathematical roles.

One of the most remarkable and specialized transformation matrices is in our brains and is a gift of nature. It takes purely physical and chemical information of the type that one finds in physics and chemical texts, and which arrive at our various sense organs, and converts it into a form in our brains that will be useful to us in everyday life. The arriving information may involve such matters as the level of the ambient temperature, or the intensity and the wavelengths of the components of light that emerge from something we are looking at. In other cases it may relate to sound waves of various frequencies. Or, the arriving information may be tied to the composition and structure of molecules which pass by the olfactory equipment in our nose. The various types of primary information mentioned in these examples, which could mean a great deal to the trained scientist, would have meant very little to our species when we were in earlier stages of evolution. The brain matrix ties them into our inner world.

1. Light

Take light, for example. The transforming matrix in the brain changes the physical parameters of the light that arrives at the eye into the sensation of color to which we are exceedingly sensitive. Thus, light with the wavelength 0.000062 centimeters is translated to the sense of *red*. The light of wavelength 0.000057 centimeters to *yellow*, 0.000047 centimeters to *blue* and so forth. Here, in fact, the matrix can play its role in a very delicate way since most individuals can distinguish differences between the colors associated with many different wavelengths or combinations of them. Indeed, such discrimination can extend to the hundreds of different shades among experts.

The main point to keep in mind here is that the colors are not "real." They are generated in our brain through the action of the transforming matrix. They gave primitive mankind the ability to sense some differences between objects out there in what the scientists call the "real" world. We cannot prove that red, yellow or blue is the same to most people—leaving aside the fact that there are various kinds of color blindness—although we suspect that the similarities are close. Almost all individuals experience a sense of excitement and cheerfulness when viewing red, of brightness and clarity for yellow, of calmness for green and depression for blue.

You might think that in view of our intense sensitivity to the various types of color we would undergo some really cosmic experience when we view all the colors of the visual spectrum—the colors of the rainbow—together and mixed. The result, however, is dead white, the equivalent of plain vanilla. This result is perhaps not too surprising since this combination is just that of ordinary sunlight which early man experienced in everyday life. He could not afford to get an emotional jag from it. He could, however, experience a sense of wonderment and awe when a ray of sunshine encountering an occasional rainstorm is separated into the component colors to form the beautiful arc of a rainbow.

One of the wonderful things the matrix can do is take the separate images from our two eyes which are separated by a small distance and fuse them so that we obtain a sense of three dimensions (stereo vision) for a range of approximately a hundred feet—a very useful facility.

2. Sound

The brain matrix seems to handle sound in a somewhat more primitive way although research indicates that the auditory system, taken as a whole, is essentially as complex and sophisticated in its own way as the visual system. Most individuals can distinguish small differences in tone. Some actually have an absolute sense of pitch. The individuals who study the generation of electric currents in the brain by observing the associated magnetic field, have found that there is a region in the brain which has a structure resembling something like a harp. Each wavelength of sound apparently produces a current in the equivalent of one of the harp strings.

Different combinations of sound, like those of light, can affect the chemistry of the brain and induce what we call a psychological mood. In classical Western music at least the minor scales induce a sense of sadness and depression.

Our two ears, like our two eyes, have stereo ability and can join to give us a sense of direction of the origin of the sound. In fact, in some as not yet understood way, they permit us to tell whether the source is in front of or behind us.

3. Odor

In many ways the most dramatic of the senses is that of smell since the olfactory organs can transform what the chemist regards as very complex chemical information, involving the composition and structure of molecules, into strong and varied signals in the brain from the repulsive to the subtly and delightfully exotic. It can also stir up memories of past events. Not least notable is the high level of sensitivity of this system, a very few molecules evoking a sensation in some cases.

It appears that the nerves in the olfactory organ in our brain, as in that of other mammals, are direct extensions of the brain as though there was a period of evolution of mammals in which the sense of smell was of primary importance. Dogs and some other animals, indeed, seem to retain much of this.

4. Generality

Some properties of the transforming matrix in our brains are clearly universal to all members of our species but others are flexible and adaptive. A given picture or scene or a given series of chords or odors may conjure up strong reactions in one individual, based on some experience in the past, and leave another individual completely unmoved. The author recalls the widow who could not bear to hear the Beethoven Quartets because she and her late husband had played them with a group in their early married years.

5. The "Real World"

Why does one bother to call attention to the transforming matrix of the brain and emphasize it in what may seem to some to be an artificially forced way? The answer is simple. Without the extensions to our knowledge of the world about us provided by the methods of scientific research, our concept of what we regard as the "real" world would be determined in the main by the kind of information which the unaided brain matrix permits to be produced in our brain. This limitation is indeed severe. One of the great strengths and beauties of science rests on the fact that while unavoidably preserving the core of the human world in us—the world to which the brain matrix is linked inside us—science permits us to gain a partial understanding of that great world out there for which our senses would normally be entirely inadequate. In effect, the methods of science with their special instruments, concepts and theories work their way both around and through the brain matrix to unveil worlds that would be beyond us. Thus we are able to "see" in the ultraviolet and infrared and to judge the distance of stars billions of light years away from us with the use of scientific equipment that brings such information within the range of the capabilities of the matrix.

While there are many areas which the brain matrix would not normally make known to us, there are two deserving special attention.

6. Relativity

The matrix was not designed by nature on the assumption that we would ever be involved in any direct way with objects travelling near the speed of light. As a result, the matrix makes it natural for us to regard time and space as things completely apart. Thus we normally accept it as natural to believe that *now* is the same everywhere. The experience gained through the outreach permitted to us through science shows, however, that for objects travelling near the speed of light, space and time become intertwined and the concept of *now* requires very special handling. All this creates what the normal brain regards as paradoxes. One of these relates to the twin who stays home while his brother travels into space at great speed and returns much younger than the stay-at-home who has aged "normally" in the meantime. This concept has been formalized beautifully in the theory of relativity and tested for precision.

7. Wave Versus Particle Dilemma

Another major area in which the natural limitations of the brain matrix show up very dramatically is in the world having the dimensions of atoms and its constituents—an area with which the brain matrix had no reason to be concerned as long as we were dealing with ordinary dimensions and with bodies for which units of feet or meters and pounds or kilograms are the natural ones. When we reach the atomic levels, entities such as electrons which orbit atoms, and which we "naturally" thought of as particles somewhat analogous to mini planets, turn out to have wave properties in addition. It is beyond the capacity of the normal mind, as a result of the way in which the brain matrix feeds information to it, to imagine an object having the properties of both a particle and a wave. A golf ball is a golf ball and a wave on the water is just that. The two do not share these properties in common, granting that both have energy associated with them while in motion.

Incidentally, the wave-particle dilemma, as it is called, first appeared while studying the properties of light early in the century but the universality of the dualism was finally appreciated in the 1920's.

The situation turns out to be even more paradoxical than these statements suggest. If you direct a beam of electrons or light particles (photons) of a single wavelength toward a baffle having two identical holes separated by a short distance, there is no way of telling through which hole the particle passes in getting to the other side if both holes are open. The most you can do in such cases is determine a relative probability, which would be 50-50 if the two holes are identical. This issue is closely linked to the so-called Heisenberg uncertainty principle which states that it is not possible to measure with arbitrary accuracy *at the same time* both the speed and position of a particle.

These major paradoxes, which have arisen from well-defined and well-established physics research, are so much in contradiction with the type of "common" sense knowledge that we derive from the everyday actions of the brain matrix that many good scientists spend their time devising experiments to test the underlying theory even though it is now nearly seventy years old. So far the theory and associated experiments seem to be moving continuously to ever firmer and indeed to very firm ground.

Incidentally, the mathematicians have devised frameworks, involving mathematical transformation matrices, within which descriptions of both the wave and particle aspects can be linked. As is often the case, mathematics can be used to express in symbolic form relationships for which ordinary language is inadequate.

8. A Complex World

One major conclusion we draw from all this is that the world out there is more complex and more subtle than the transformation matrix in our brain permits us to appreciate unless we carry out research which permits us to extend our reach far beyond its limitations.

There is also a complementary side to the story which is fully as important. Whatever we learn from the broad pursuit of science must, in one way or another, pass through the gateway controlled by the brain matrix and be reduced to concepts that we can relate on our inner human terms, encountering paradoxes if we must. After all, the only language we have is the language of our own minds, particularly the language we share in common. This means that we must not only expect to meet additional paradoxes as science proceeds but we must never mistake the artifacts and concepts that the matrix permits us to develop in our brain for the way in which things "really" are out there. Or as the great philosopher-scientist, Immanuel Kant, expressed it, we can never really know in any complete sense the true or inner essence of things as they really are in the world about us.

2
Technology Without Modern Science

The engineer
is a remarkable man.
He knows little science
but create he can!

(Free translation of a 19th century jingle; courtesy of Professor James Franck)

Our species, in addition to exhibiting many other remarkable attributes, particularly the capacity to develop very complex languages and social systems, is a maker of tools and a designer of systems of tools. Our primary equipment for this task consists of manual dexterity, stereoscopic vision, sufficient intelligence to carry through a degree of logical reasoning, sufficient patience to become involved in trial and error enterprises, well-developed means of communicating with one another, and the ability to transmit a degree of acquired knowledge from one generation to the next. We emerged on the scene with these aptitudes in fairly developed form about fifty thousand years ago and by using them have essentially taken possession of the planet in the intervening period. Our ancestral hominid species possessed some of these aptitudes but none to the degree exhibited by homo sapiens. Some time in the next century we could, if we chose, take possession of the solar system. That, however, will depend upon social, psychological and economic factors.

This advance, while basically continuous, has had well-defined mileposts. For example, about ten thousand years ago we became agronomists rather than primarily food gatherers and hunters. Our antecedents moved their agriculture into the great river valleys such as the Nile, the Indus, the Euphrates and the Yellow Rivers about five thousand years ago and, among other things, used the renewable silt associated with annual floods to maintain fertility. The high yields of river valley agriculture per farm worker made it possible to develop a sophisticated urban society and the specialization of classes that go with it. Many modern professions were born in those societies including what is today termed engineering. In fact, all fields of modern engineering except electrical and nuclear engineering can be clearly recognized in a basic form in these earlier civilizations.

1. Ancient Recognition

In Greek mythology the god of ingenious innovation is Hephaistos, a worker of wondrous skills. He is characteristically regarded to be lame, presumably to afford an explanation of the fact that his genius is better suited to occupy a place at the forge or in the workshop than in the front line of battle. As a reward for his creativity, he is wedded to Aphrodite, the goddess of beauty and art. This union carries with it the symbolic implication that the Greeks appreciated the fact that the creativity of the artist and that of the ingenious technologist are closely linked. Incidentally, it appears that the name Hephaistos is of semitic origin, implying that the explicit veneration of technology and its fruits appeared in the earlier semitic civilizations in the Near East and Mesopotamia. In the Roman pantheon, Vulcan and Venus held positions equivalent to those of the god and goddess of the Greeks.

2. Northern Europe

When the previously nomadic tribes of northern Europe settled into an agricultural existence they faced soil and climatic conditions quite different from those encountered in the Mediterranean and Near East. Through trial and error, they instituted a significant revolution about twelve hundred years ago by learning to deal with such conditions. Their success depended upon the development of deep cutting, iron-reinforced plows, learning to equip a horse to draw such plows efficiently and evolving a pattern of crops and crop rotation that would yield sufficient excess food to permit the development of an urban society well north of the Alps. Without these great technical advances the discovery of North America or of the temperate zones of South America would have been of relatively little importance to the Europeans beyond the discovery of new mineral wealth to be found in these regions. It might be added that these innovations probably had much to do with the ultimate settling down of the Scandinavian people in the 12th and 13th centuries after the period in which the Vikings were the terror of Europe.

3. Common Pattern

The pattern of technology that is based on common sense reasoning and trial and error, coupled with inventive genius, which together made advances in civilization possible, permitted many other important innovations. Contained in them are advances in the development of various implements of war including fire arms, many of the most basic aspects of ferrous and non-ferrous metallurgy, the evolution of mining operations, the development of the sailing ships which made the Age of Exploration possible, the invention of precise mechanical clocks, the development of printing, and the discovery and refinement of countless compounds both organic and inorganic which are used by all of us in everyday life or by specialized artisans. In fact the basic techniques which produced such innovations, depending as they did upon accidental discoveries incident to work or play, trial and error, and intelligent observation, carried society well into the age of coal and steam.

It seems safe to say that almost all great technological advances of civilizations prior to modern European civilization, and, in fact, practically all the advances in Europe to about 1800 with one or two notable exceptions depended upon continued ingenious application of the basic methods which our species has been using quite effectively in one way or another for approximately fifty thousand years— techniques which one may term those of classical technology. In this pattern, opportunity, necessity and invention worked in unison to inspire the practical mind. In truth, we recognize in the methods of classical technology the techniques we all use in an amateur way to resolve countless problems of everyday life, whether handling personal or family matters or participating more generally in the affairs of society. Small children use them in their own way to solve their own special problems.

The ability of our species to pursue what we have termed classical technology is evidently an intimate part of our birthright—it is closely linked to the genetic makeup of the average member of our species since it is found everywhere and has appeared in profusion whenever social conditions have fostered it. It is true that the aptitude for doing highly creative work by the methods of classical technology varies from one individual to another. However, it is scarcely ever absent from any community of human beings. It is also true that there are societies which have frowned on technological advance from time to time because they fear that innovations might have a disruptive effect on their social institutions or undermine the pattern of traditions which regulate their society. Such restraints have had to be imposed formally, however, since it appears that the members of our species are innately inventive and will do their best to exercise this talent when free.

Some modern engineers often object when the methods that make classical technology possible are called the techniques of "science". Perhaps they have a point. It might be mentioned, however, that the word "engineering" is not much older than the word "science." In the United States, for example, the word "engineer" as applied to civilians has been in common usage for only about one hundred and fifty years. It follows that this celebrated semantic battle of our time is perhaps something in the nature of a tempest in a teacup.

4. The Rise of Basic Science

The evolution of the web-work of basic science, that is, the pursuit of science to obtain deeper knowledge of nature, will be discussed in the next chapter and need not be dwelt upon in any significant detail here. What is important is that the rise of basic science is a relatively late comer to the human scene. It could emerge and develop only after civilization had progressed to a stage where individuals could exercise their innate curiosity concerning nature freely and had both the freedom and the inclination to do so. From its start in the special climate provided by the Athenian state for about two and a half centuries, it tended to have a life of its own and to flourish differently in different civilizations.

One could not have had the birth of science without significant prior advances in technology. It so happens that all the major civilizations leading up to the Greek had need for land measurement and accounting and thereby developed some form of mathematics which varied in sophistication depending on the civilization. All such civilizations tended to use the cycles of celestial changes to guide agriculture and to time religious exercises within the framework of their own local conditions. Without this base, any form of speculation that the Greeks had engaged in during the fruitful centuries would have had little solid content or permanent value.

It will be emphasized in Chapter 3 that we owe a debt first to the Romans for not interfering with the preservation of the early Greek work or with its subsequent extensions in the early Christian period and second to the Arabs for their very fruitful synthesis of all available technical and scientific information to be found within the reaches of their empire, which extended from India to Spain. Inasmuch as the Islamic culture was highly literate, their scholars expounded and expanded upon this material in such a way as to continue to give it permanent form.

5. Transfer to Europe

The transfer of the Arab scientific literature to Western Europe ignited a very special flame which placed science on what proved to be a new and even more fruitful basis. The Christian European scholars not only engaged in relatively uninhibited speculation but insisted in a steadfast manner that while speculation could roam freely it must be tested by specially designed experiments if need be. This principle, coupled with the application of rigid standards of logic and the continued evolution of mathematics, made it possible to develop a wide framework of new insights into the world of nature, in brief into the world of what we term modern science.

To repeat, science had, at first, little impact on technology. Probably the first significant application among the Europeans was the help the 15th century Portuguese navigators derived from the early Greek estimates of the diameter of the earth. There is no doubt whatever, that Henry the Navigator's scholars at Sagres clearly understood that the earth was a sphere. One might also wonder if they also appreciated in a practical matter-of-fact way such matters as the diurnal rotation of the earth and its annual course about the sun decades before Copernicus published his great treatise. The Portuguese had access to the thoughts of Heracleides and Aristarchus, as well as Nicholas of Cusa, whom we shall discuss in the next chapter, and presumably to some of the practical experience of the Arabs. Unfortunately, both the veil of secrecy maintained by Henry's university and the subsequent accidental destruction of records prevent us from knowing more.

As an aside, it may be mentioned that we know even less regarding the scientific knowledge of the Phoenicians who were contemporary with the Romans in the pre-Christian era. According to Herodotus they were sufficiently enterprising to circumnavigate Africa and noted that in the southern hemisphere the sun traverses the heavens in the north rather than the south. This must have set the minds of these imaginative peoples, who undoubtedly knew the details of Greek speculations, into contemplative channels of their own. Unfortunately their secrecy, combined with the destruction of Carthage by the Romans in 146 B.C., leaves us with another historical mystery.

6. Wholesale Application of Science

Generally speaking, we must wait essentially until the 19th century to see the wholesale application of scientific knowledge to engineering. It is, however, worth noting that a few of the leading engineers of the renaissance period appreciated the significance of the advances in science and were inspired by them. For example, Simon Stevin, the Flemish engineer who helped to evolve the canal system of the Netherlands in the 16th century and was the architect of the most modern fortifications of the time, was sufficiently inquisitive and imaginative to discover the law of the triangle of statically balanced forces as well as the principles of hydrostatic force at various depths. Many of the most brilliant and flexible minds of the period were capable of combining an appreciation of the practical and the philosophical to the advantage of both. In the main, however, the engineers of the day proceeded serenely on their way. Most of the artisans probably respected science from afar but used their judgments and skills along traditional channels in carrying through their everyday work. The true wedding of science and engineering was still several centuries off.

7. The Question

This takes us to the heart of the present discussion and leads to the question "Where would technology be today if the scientific method had not been developed?" What might have happened if the western Europeans had been like the Romans and devoted their talents almost exclusively to everyday practical affairs, pushing ahead along traditional lines much as the Romans pursued military and civil engineering without devoting very much attention to the philosophical ideas of the Greeks?

One must admit that the question is a rather artificial one in many respects because the universal ferment found in Renaissance Europe had precisely the characteristic that it inspired the scholar and the artist as well as the merchant and the engineer. All were gripped in their own way by an all-pervading spirit of new enterprise. Nevertheless, the question has considerable meaning, just as it might be reasonable to ask what the plants and animals of the Americas would be like if there had never been a Siberian land gap or if our continent had never made successive contacts with Europe and Africa as a result of continental drift.

8. The Situation Before 1800

As mentioned earlier, matters up to the year 1800 probably would have continued more or less as they had if science had not developed or if it had remained isolated from technology. Most of the products of classical ceramics and metallurgy would have come into use. Coal and steam would have been exploited for heat and power. The age of geographic exploration would have gone on almost unchanged and would have stimulated the development of nautical instruments including the chronometer. Without prior knowledge of the conclusions drawn by the Greeks concerning the shape of the earth, and transmitted by the Arabs with their own experience south of the equator, the early Portuguese navigators who ventured down the west coast of Africa might have been surprised to see the sun moving to the north and the polar star disappear completely. Nevertheless, they would have persevered in their course since it had practical economic goals. The age of petroleum would have blossomed much as it did in the middle of the last century and we would have had the automobile and mechanized farming equipment. Clearly the internal combustion engine would probably have taken on a different aspect because of the absence of electrical ignition but we still would have had diesel-like engines and gas turbines, perhaps using self-igniting fuels or other means of ignition. Indeed, the diesel engine might have occupied a much more central position and undergone far more ingenious innovation.

On the side of the life sciences, plant and animal breeding should have continued to progress along traditional lines. Medicine should have evolved, probably with the discovery of more elementary anesthetics than are used at present. The life sciences, however, as in the past century would have been clouded in mysticism. Quasi-mystical approaches to the technology of living systems would have been a common phenomenon everywhere as they still are in the non-scientific literature of today.

We would have been far more ignorant of natural phenomena even in the large. However the Age of Exploration should eventually have demonstrated to one and all that the earth actually is a globe. This practical proof of the form of the earth might have led many to suspect that the planets moved about the sun rather than about the earth as had been suggested by Aristarchus. However, everyday technology would have been affected only indirectly by such revelations, deriving from them inspiration more than immediate tangible benefits.

It should be added that eyeglasses, the telescope and the microscope could have come along in at least rudimentary practical form following the routes of traditional technology. After all, the spectacle and the telescope were devised well before optics became much of a science. Kepler was the first to give what might be called a modern theory of the telescope. It, however, was proffered after the invention.

9. Breakthrough in Chemical Technology

The first really significant omission which would have had a major influence on technology is the science of chemistry. While the alchemist, who was scarcely scientific in the modern sense, and his more immediate colleagues, the forerunners of the chemical engineer, would have isolated hundreds of useful compounds over the centuries using relatively crude trial and error techniques, the great advances in technology, including medicine, would have been hampered very significantly in the 19th century without access to the growing science of chemistry. The knowledge of inorganic chemistry would have been highly fragmentary and would have been clouded with the type of mysticism the alchemist revelled in. The coal tar industry could scarcely have been born. Relatively sophisticated compounds such as trinitrotoluene and nitrocellulose might have been unknown, although black powder would still have been available to provide the explosives for great civil works and great wars.

In brief, many aspects of modern living would resemble somewhat those familiar in Europe and the United States in about 1875 with the possibility of one great exception, namely the continued evolution of sophisticated mechanical gear such as the gasoline engine. Since farm machinery, the automobile and possibly the airplane could have emerged as a result, urban society would have continued to grow. Transportation could have been at least as rapid as it was in the first half of this century even though the standard of living of the Atlantic community would have been somewhat more primitive. It seems safe to say, however, that without the evolution of chemistry, the technologist would have been hitting a ceiling by the present time because the materials available to him did not evolve nearly as rapidly as his needs. Mankind would be facing a limitation of resources.

10. Technology with Chemistry as the Only Science

We might play our exercise somewhat differently and admit the development of the more qualitative sciences—more specifically chemistry—into the picture. Even without the drive provided by the explicit formulation of the principles of the modern scientific method, it is not unreasonable to assume that in the 18th century highly imaginative individuals such as Priestly or Lavoisier would have caught on to the concept of the chemical elements and, with their associates, started to unravel the composition of matter, much as actually occurred. This innovation, after overcoming many obstacles derived from the quasi-mystical past of the field, should have made it possible to push ahead with the development of more qualitative aspects of structural and analytical chemistry along the lines followed by Dalton and Kekule and their successors. This in turn could have led into the heart of the age of synthetic chemistry and provided the technologists with a host of inorganic, organic and biological compounds so important for 20th century engineering.

What might be termed crystal technology would have gone fairly far in the scheme proposed here since mineralogy as a branch of geology has done much to provide classification. The existence of a lattice structure would probably have been inferred by chemists. Some good guesses about elementary structures might have been made as indeed was the case for a simple material such as rock salt early in the present century before x-ray diffraction was discovered. It is evident, however, that progress beyond the most elementary cases would have been very limited and the lattice structures of most compounds remained a mystery.

It follows that armed with chemistry the engineer, agronomist and medical technologist provide us with a very large fraction of the benefits we enjoy in this century. Moreover the prospects for further advance would still be fairly bright.

11. The Science of Electromagnetism

The second great omission we would have experienced, almost as serious as that faced without the science of chemistry, is the science of electromagnetic phenomena. It is exceedingly doubtful if classical technologists could ever have stumbled into the heart of this field by using purely trial and error methods as they might well have stumbled into parts of modern chemistry. For one thing, it required unusual inquisitiveness to pursue the development of scientific curiosities such as charged pith balls, the voltaic cell, and the electrostatic machine. Without such endeavors and the evolution of associated instrumentation, initially of purely scientific interest, most of the investigations that lead to the basic equations of electromagnetism would have been missed. Traversing the long path that leads to the means of generating such things as radio waves was probably far beyond the patience of even the most doughty engineer. The special qualities of the research scientist were required. We would have been deprived of electromagnetic machinery as well as knowledge of electromagnetic waves.

It is true that if electromagnetic phenomena had not been discovered, much of the genius which was devoted to the application of electromagnetism would have been focussed on mechanical devices so that they should have been pushed to the ultimate as was suggested somewhat earlier. The gains associated with such a shift in emphasis are far from trivial. However, we would not have had dynamos and motors for power generation and conversion. Moreover, it would have been necessary to produce and use power locally so that the pattern of power distribution should have been quite different from that which we have today, although we would have had extensive networks of pipeline for oil and gas. Still further, we would have been deprived of the essentially instantaneous communications made possible by telephone and wireless. No mechanical system can be expected to operate at anything comparable to the speed of light, granting that line of sight optical signaling might have evolved to a degree. The so-called "gay 1890's" which witnessed so many innovations in electromagnetic devices and gave society an encouraging glimpse of the technical potentialities of the 20th century would have been much less exuberant. We should, in short, not have had an age of electricity.

With the world of electromagnetism closed off from us the world of the atom would have been much more difficult to explore even though the chemists, whom we have allowed to survive and in fact to thrive, could have surmised the existence of the atom in a relatively qualitatively and descriptive way following the lines of Dalton and his more quantitative successors who worked out chemical formulae. Not only would phenomena associated with gaseous electricity have remained undisclosed, but electronics and x-ray technology would have been unknown. In brief, all the great inventions of the 20th century that were derived from physics would be absent. The computer should have remained a mechanical device and the great new world derived from the use of integrated circuits would have been almost beyond imagination.

12. The Life Sciences; Public Health

In the area of the life sciences semi-practical studies of plant breeding along the lines of Mendel and his successors might have given some clues concerning the existence of genes. However, without the full range of tools provided to the cellular and molecular biologists through the channels of physics, such as the high speed centrifuge and the electron microscope, knowledge of the inner workings of the cell would have been very limited, although the fundamental experiment which demonstrated that DNA is the carrier of the genetic message might still have been possible if we are prepared to be sufficiently imaginative. Taken as a whole, however, knowledge in the field of biology would have been very closely tied to the practical advances in agriculture and medicine.

Without chemistry, public health would have improved through the work of the mechanical, sanitary and agricultural engineers who have done much to advance our welfare. However, we could not have expected to go much above the best standards of the third quarter of the nineteenth century. With chemistry, most aspects of public health should be comparable to those we experienced up to about 1950, *if* we permit the addition of vaccines and antibiotics. Treatment of the more subtle virus and other cellular diseases would probably have been beyond reach. While they still are beyond reach to a significant degree, we do have a solid basis for hope.

13. Nuclear Science

It is faintly possible that chemists, working with luminescent compounds or photographic plates might have discovered some of the remarkable properties of radioactive compounds. However, it is very doubtful that this would have led to the revelation of the nuclear particles or the structure of the nucleus with anything resembling the quantitative clarity made possible with the use of electromagnetic devices such as magnetic deflectors, ionization chambers and the like. The neutron would probably have remained undiscovered. It is scarcely imaginable that a classical technologist could have ever dreamt of assembling the array of fissionable materials and moderator which appears in a simple fission reactor. In brief, the whole world of nuclear energy would have been closed off to mankind.

14. Enlightenment

One aspect of the arrival of science on the human scene which is very difficult for us to evaluate at this stage of the evolution of technology, and indeed may always be elusive, is the profound effect it has had in providing the special form of enlightenment which makes it possible for the innovator to move ahead relatively free of the shackles engendered by ignorance, superstition and mysticism so prevalent in earlier times. Science has provided us with a new light with which to view the world in which we live and to illuminate the path along which we travel. Perhaps this is the greatest gift of science to technology. Indeed, the revelations of science undoubtedly encouraged the development of new attitudes in many other areas of human endeavor.

What do we conclude from all this? Without the benefits of science the classical technologist might, in principle at least, have been able to push ahead in a remarkable way well into this century giving us many of the things we profit from today, perhaps including the airplane, with appropriate perfection of the steam engine or its equivalent and the diesel engine. He would, however, at some time in the mid-20th century have begun to encounter increasing constraints of a kind which still lie far in the future for present day society.

If we grant him an alliance with classical chemistry, emerging from the discovery of the elements, he would have been able to push along faster and with more sophistication, but he would have had none of the benefits of electromagnetic or electronic devices and would, by the end of this century, begin to be seriously cramped by a reliance on fossil fuels.

Viewing the situation as a whole, it is clear that the twentieth century is the one in which the full importance of science for human welfare really became evident. In the next century the continued progress of our species will be almost wholly dependent on innovations which emerge out of the scientific revolution.

3
Crucial Steps in the Evolution of Science[1]

> The inquisitive spider that is called science weaves a web of its own choosing. Its ultimate pattern is never really mysterious but almost always unpredictable.

Albert Einstein's friend, J. E. Switzer, asked him why the Chinese had not invented or discovered what we term modern science in spite of their obviously great creative talents. Einstein's response, probably in part jocular, was to the effect that it is a miracle that modern science evolved anywhere.

Without expressing anything but unqualified admiration for one of the greatest scientists of all time, I do believe the question deserved a more thoroughgoing answer because it is clear that the successful pursuit of science is by no means out of keeping with the capacities of our species. In our own day, the creative scientist, while an unusually endowed individual, is not a basically abnormal human being. He or she has characteristics of the type that can be found in some individuals in any developed society. Let us return, however, to the Chinese issue later and discuss the factors that finally permitted the successful arrival of what we call science on the human stage.

First, consider the principal factors required if science is to move ahead in any society. The following list includes eight items which are basic:

---

[1] The writer, a scientist, is not a professional historian. The material presented here is derived from secondary sources which are on the whole in agreement with one another. Some of the sources used are listed in the bibliography at the end of this chapter.

This article is an abbreviated version of a manuscript prepared for a conference in London in August, 1989 on the subject "Liberal Democratic Societies: Their Present State and Their Future Prospects."

He is grateful to numerous colleagues for comments and particularly grateful to Professor Subrahmanyan Chandrasekhar of the University of Chicago and to Professor A. I. Sabra of Harvard University. The former provided him with a volume on the history of Indian science published by the Indian Academy of Sciences whereas the latter provided him with seven highly informative essays on major aspects of Islamic science.

1) The society must be well beyond the hunter-gatherer stage and far enough along in the evolution of agriculture and animal husbandry that it permits what might be called leisure and professional classes. That is, it must be civilized and possess significant centers of culture. We have had many of these in the past five thousand years or so, probably starting with the Egyptian civilization which grew out of the intrinsic wealth of the Nile.

2) The society must contain a nominal number of individuals having the freedom to ponder, from curiosity and their own free will, the wonders of the world about us. This characteristic appears to be omnipresent although it may be expressed in different ways in different cultures.

3) A reasonable fraction of those individuals who are interested in such pondering must have the logistic support needed to permit them to do so. In the early phases of the development of science their requirements will be modest beyond the essentials of life. The Greek civilization of which more will be said later, must have supported quite a few thousand individuals of this type over a period of a thousand years or so.

4) There must be readily accessible institutional frameworks that permit the contributors in the search for knowledge to live in an environment that is reasonably compatible with their needs. In our own day we have universities, research institutes both private and public, and distinguished industrial laboratories.

5) There must be systems of communication and information storage which permit the investigators to interact as broadly as possible in order to assure an essential exchange of ideas over a period of time, both to provide criticism and to permit evolutionary development.

6) There must be great pressure within the evolving system to promote, and indeed insist upon, what might be termed logical analysis of conclusions which are drawn by individuals. Such conclusions are all carried out with the understanding that the community of investigators will strive to reach common agreement on principles that are valid as common working tools, at least for a period of time until a logical reason for change, arising out of science itself, may appear as part of the evolutionary process. Observed deviations from commonly accepted conclusions must form the basis for careful reexamination of such conclusions.

7) There must be a strong interest in experimentation that permits ever deeper and more thorough understanding of the material (physical) world. Greek science ultimately foundered in significant part because of the failure to appreciate this basic requirement.

8) Finally, those involved in the pursuit of science must be substantially free from dogmatic pressures which impede their freedom of speculation and experimentation. Such pressures can arise from forces within the social structure, external to the group of investigators, and deflect them forcibly from their activities, or they may be imbedded within individuals as part of their indigenous cultural heritage

in a form that is essentially impossible to overcome. In lieu of better evidence than we have at present, one suspects that the second factor impeded the development of science in China.

Let us now sketch the pathway to modern science.

Step 1—The Greeks

Greek Science

We begin with the Greeks both because of the great influence of their work on the rise of science in Western Europe in the last seven hundred years and the existence of reasonably clear documentation of Greek science, granting that some of the knowledge the Greeks possessed and transmitted, particularly aspects of mathematics and astronomy, was probably derived from other civilizations, particularly those of Mesopotamia, Persia and India, both earlier and contemporary, by diffusion if not directly.

The Greeks arrived in the Aegean area from the north during the second millennium B.C. and soon evolved a highly significant culture distinguished in several ways. Although they were reasonably homogeneous linguistically and ethnically, the geography of the Aegean led to the development of numerous politically independent groups. They communicated freely at the cultural level but in a disputatious way in which logic and reason were used as tools and weapons as circumstances required. Moreover, their curiosity concerning the workings of the natural world was essentially unlimited.

The first great school of Greek science, termed the Pythagorean, was established about 500 B.C. and focused on, among other things, the field of geometry. This aspect of its heritage is recorded in part in the books of Euclid (320-260 B.C.). The development of a fine-tuned appreciation of geometry made a deep imprint on the Greek mind and had a profound influence on the approach the Greeks made to the physical as well as the mathematical sciences.

There seems to be no doubt that the theorem of Pythagoras governing the relationship between the sides of a right angle triangle was appreciated and used in special cases, such as when the sides are in the ratio 3:4:5, long before the Greeks, particularly by the Egyptians, Mesopotamians and Indians. This, however, does not detract from the majesty of the work of this school.

The Pythagoreans decided that the earth is spherical, probably from observing the passage of its shadow on the moon during a lunar eclipse.

Two of the towering Greek intellectuals were, of course, Plato (427-347 B.C.) and Aristotle (384-322 B.C.) both of whom roamed over the entire area of human culture within their horizons. They overlapped in time, the latter spending a number of years in the academy founded by the former. In his younger days Plato expressed a not uncommonly held view that the results obtained by hands-on experimentation were not to be trusted relative to the speculations of the unrestrained human mind which he regarded as more refined and more trustworthy. While he apparently recanted in later life, hard-core experimentation, except for astronomical observations, did not become an underpinning of Greek science at its most critical stage for one reason or another. We must, however, give great credit to Archimedes (287-212 B.C.) as a notable exception. He experimented with levers and with bodies immersed in liquids as well as many other devices, and gave us the well-known laws governing them. He was also active in the defense of Syracuse and contributed substantially to the military technology of his day. He may indeed be regarded as a speculative engineer.

Aristotle, who has the distinction of being the tutor of Alexander the Great, cast a very long shadow on the history of science. Viewed from the modern standpoint, he was probably at his best in the field of biology since he initiated a classification of species which was not unreasonable for his time. However, he also took on the systematic formulation of the basic laws of physics and produced a theoretical system which had a profound influence on the history of science for essentially two thousand years—an influence which was both positive and negative.

In brief, and to use modern terminology, Aristotle postulated that as a dynamical system the physical universe possessed an absolute reference framework centered at the core of the earth which is assumed to be stationary in space. The natural state of any physical body on the surface of the earth that is not subject to an external force causing it to move is to be at rest—a state which it will assume as soon as any applied force is removed. Moreover, a falling body would, if not impeded, come to rest at the center of the reference frame in the earth, its natural home. This system is obviously in accord with what one might term everyday observations if one does not go too far afield in testing it. Stellar objects were assumed to be made of material quite different from that on earth and to be governed by their own special rules, revolving about the earth in divinely ordained circles.

One observation that did bother Aristotle was the continued flight of the arrow after it leaves the bowstring. He resolved this issue by assuming that convective air currents generated from the original push of the bowstring kept the arrow on its way for a relatively short time before it dropped vertically. He also said that the gods could not keep an arrow in flight in a vacuum.

Two other items of classical Greek science should be noted here. First, Eratosthenes (276-194 B.C.) determined the radius of the earth to a remarkable degree of accuracy by observing the angle of the sun at two latitudes on the same day of the year. Second, Aristarchus (310-230 B.C.) proposed, on geometrical and aesthetic grounds, that the planetary system is solarcentric rather than geocentric and that the earth rotates on its axis once each day. He did not receive a sympathetic response from either his intellectual colleagues or the larger public so the concept was not built into the framework of Greek science even though some astronomers clearly recognized that the inner planets, Mercury and Venus, do circle the sun.

In a sense, the ground for Aristarchus was broken by a predecessor, namely, Heracleides (Ponticus), who lived between 390 and 322 B.C. He had the boldness to suggest that it would make good sense to assume that the stars were fixed and the earth had a daily rotation about an axis. He also proposed that Mercury and

Venus rotate about the sun. It is not known whether he also attempted to propose that the earth and the other planets rotate about the sun since most of his writings are lost. For ideas such as this, he gained the label "The Paradoxologue." Either the Greek credibility or its willingness to be flexible in the face of more conventional views was being stretched to the limit.

Archimedes, who overlapped with Aristarchus, was well informed regarding his theory of a solarcentric solar system and reported it to King Galon, his patron. One can only guess how much credence Archimedes gave to the proposal.

A number of estimates of the distance between the earth and the moon were made on the basis of the determination of the relative sizes of the earth and moon from projections of the shadow of the former on the latter during lunar eclipses but with variable results, usually less than the true distance. Estimates of the distance to the sun were even less accurate, but it became clear that the sun is larger than the earth.

Step 2—The Roman Period

Influence of Greek Science in India

It seems probable that Alexander's expedition to India (326-327 B.C.) provided the Indian scholars, who had a great interest in mathematics and astronomy since ancient times, with some appreciation of the Greek work in geometry. It is not evident, however, that the very important Indian advances in analytical aspects of mathematics had a significant *direct* influence on the West until Islamic times nearly a thousand years later when Baghdad became a great cultural center and a new synthesis occurred.

Ptolemaic and Roman Times

Following Alexander's untimely death in 323 B.C., after his conquest of Persia and Egypt, and the expedition to India, his empire was divided among three generals. Egypt fell under the control of the Ptolemys who governed it until the arrival of the Romans at the time of Julius Caesar in the 1st century B.C.

In the meantime, the city of Alexandria in the Egyptian delta with its famous lighthouse and library became one of the great cultural and study centers of the world. Many Jewish merchants and intellectuals moved there after the diaspora which followed the War of 70 A.D. Indeed they played a significant role in the stimulation of life throughout the Mediterranean World.

The Romans did not interfere in any direct way with the further development or dissemination of Greek science for which they had much respect. The medical sciences continued to receive a great deal of active attention and the Medical Treatise of Galen (131-201 A.D.) served as one of the great authoritative sources of medical knowledge for a number of centuries, in fact until the time of Vesalius and Harvey in the 17th century.

The highly innovative period of Greek science which lasted from about 500 B.C. to about 300 B.C. and, in a very real sense reached something in the nature of a termination with the rejection of Aristarchus' solar-centered planetary system, was followed not so much by stagnation as by ferment of a very different kind from that which had taken place earlier. Philosophical speculation, much of which occurred in Alexandria, ran off in many directions involving schools devoted to Epicurean-ism, a resurgence of Stoicism and the renascence of Platonism. Part of this was retrospective in the sense that the works of Plato and Aristotle, for example, were subject to detailed scrutiny.

This phase of Greek intellectual ferment reached its peak in the era of Neoplatonism during the 3rd century of the Christian Era. One of the influential leaders in this period, namely, Plotinus (205-270 A.D.), was an Egyptian schooled in Alexandria who became a distinguished figure in Rome where he led a major school. There is good evidence to indicate that Aristotle's theories of motion were subject to critical discussion during this period. The modern historian of science, Shlomo Pines, has found an interesting statement in the writings of St. Augustine (354-430) which suggests that there was great interest in the continued movement of bodies once set in motion such as those thrown or dropped. In fact, St. Augustine introduces the term "impetus" to describe the cause of such motion. It is quite possible that Buridan, whose work will be discussed later in this chapter (Section 8), adopted the term to his own use as a result of his familiarity with the work of St. Augustine. It should be emphasized that there is no evidence that the philosophical discussions associated with the neoplatonists were accompanied by significant experiments.

It may be added that Christian doctrine was churned by the various schools of philosophy during the period up to the late Roman era when Christianity finally became the official religion of the Empire. Essentially all the philosophical formu-lations of Christianity developed prior to that were rejected as "heretical" once the Roman Catholic church became firmly established and new, more official doctrines entered the picture.

Mathematics and astronomy continued to flourish within the traditional frame-work of Greek culture. The astronomer Ptolemy, who lived in the 2nd century A.D., both extended and worked over the existing astronomical data in a desire to perfect knowledge of planetary motions, all within the assumption of a geocentric system. Following earlier suggestions he concluded that improvements would indeed be made if, instead of assuming simple circular motion about the earth, one adopted a system of epicycles. More specifically, he assumed the planets revolve in circles which are centered on the circumferences of still larger circles which in turn are centered at the earth. There is no evidence that he attempted to revive Aristarchus' notion of a solarcentered system.

At this time there was much interest in sophisticated gadgetry, such as the devices constructed by Hero of Alexandria, who lived in the 1st century A.D., which were based on relatively subtle physical principles such as jet propulsion by water or steam or the action of a siphon. The level of advance of instrumentation involved is noteworthy. Perhaps this form of invention was encouraged by the Romans who had a well developed appreciation of mechanical systems.

It is remarkable that the Romans who displayed such great inventive genius in the fields of civil and mechanical engineering did not themselves contribute a great deal to the advancement of basic creative science in spite of the substantial Greek activity taking place in their midst. One can only conclude that, in ways not yet well understood, cultural influences acting through individuals rather than any direct external restraints were responsible. As has often been said, the Romans tended to have practical minds.

Christian Rome

By the time Rome adopted Christianity in the 4th century A.D. and formed what might be termed an orthodox Christian church, great changes in values had occurred, none of which gave great stimulus to the advancement of science. However, the establishment of an official church, which was inclined to develop rigid doctrines of belief, inevitably led to the generation of dissident or heretical Christian groups which espoused contradictory doctrines. Such divisiveness has been true of all established religions wherever they may be and will lead to conflict if the leaders so choose. The concept of the Trinity, which became central to official Christian orthodoxy, was, for example, a source of much contention. In this connection the priest Arius (250-336) was banished to frontier areas for preaching, in contradiction to the principle of the Trinity, that Christ was merely a perfect mortal who attained a special spiritual status through his career and sacrifice. It is an interesting fact (Section 18, this chapter) that Isaac Newton strongly, but secretly, favored the views of Arius in spite of the fact that he held a post at Trinity College at Cambridge University.

Some of the dissident groups moved away from the centers of Roman civilization to more isolated regions in Asia Minor or beyond, taking with them the great books of learning in the original or translated versions. Some of these communities found themselves absorbed into the Arabian Empire as Mohamet's followers swept out of the Arabian Peninsula in the 7th century A.D. and took over large portions of the Roman world both east and west.

Step 3—The Arabs

The Rise of Islam

Fortunately the Arabs displayed considerable tolerance for the dispersed religious groups and their learning and reached out for advanced knowledge, not in conflict with the precepts of Mohamet, that had previously been unknown to them. They welcomed both foreign scholars and their science texts wherever they found them. Many of the books were rapidly translated into Arabic.

Both Moslem Mesopotamia and Moslem Spain soon generated great centers of what might be termed international Islamic culture. What was inherited from the Greco-Roman world as well as from Persia and India was enhanced by major additions.

A particularly remarkable phase of integration of diverse cultures and learning took place in Baghdad during the first century of the reign of the Abbasid Caliphs (750-850) when political power was highly centralized in the empire. This included the period of the reign of Harun al Rashid (786-809). During this time, Baghdad acquired a highly diverse population from all over the Near East and became the largest city in the world outside China with a population of the order of a half million persons.

It would be difficult to overstate the important role played by the multi-ethnic scholars of Islamic civilization as carriers and extenders of the scientific knowledge and traditions they inherited from the Greeks as well as other cultures. Their universities and libraries became great centers of study involving Arab, Persian, Christian, Jewish and other scholars, undoubtedly including Indian. They extended investigations in such fields as medicine including surgery, chemistry (or alchemy), astronomy, optics and mathematics. Algebra and decimal notation, first developed in India, were major additions to mathematics. Not least, they improved upon old instruments for quantitative measurement and invented new ones including much more advanced versions of the astrolabe. It is significant that the word "calibrate" so important for quantitative measurement is of Arabic origin.

In spite of the brilliance of Islamic civilization, Islamic scholars seemed to be acting under almost invisible constraints probably arising from the all-pervading influence of religious doctrines. Although the scholars were completely familiar with Aristotle's work, including his physics, it was, on the whole, not subject to intense criticism. It may be noted however that Avicenna (Ibn Sina, 980-1037) a brilliant Persian philosopher and critic, did suggest that air resistance leads to the dissipation of the movement of a body set in motion and that in the very unlikely circumstance that one could produce a vacuum it would be found that a moving body such as an arrow would continue its motion unimpeded until striking another object. He proposed that a body set in motion possesses an extra quality not present in one at rest. One must admit that in principle at least this proposal is close to the modern concept of inertia introduced by Galileo and more sophisticated than that of Buridan (Section 8). Again, there is no evidence that any systematic experiments were carried out in support of such concepts.

Another Moslem scientist Alhazen (Ibn Al-Maitham, 965-1039), contemporary with Avicenna, made great advances in geometrical optics through experiment and theory and wrote important texts on the subject.

The universe remained geocentric, however, although the framework of Ptolemy's system of epicycles was subject to criticism. Galen's medicine was expanded and used but not revolutionized, although instruments and techniques for surgery were improved.

The initial flowering of Islamic civilization occurred at a time when the Western portions of Christian Europe including Italy were struggling with the aftermath of the collapse of the Roman empire and the absorption of semi-barbaric immigrants from the east and Scandinavia. In these circumstances the local Christian scholarly institutions tended to become crude and primitive at best. Much of the knowledge gained from Greek science was lost or rejected. For example, the notion that the earth is flat became widely accepted as a part of religious belief. Medicine regressed. While there is no doubt that some of the classical traditions and documents were preserved in Byzantium, centered in Constantinople, the pursuit of science became moribund there as the society faced complex political, social and religious problems, both internal and external. Preserving a stable government and protecting outlying provinces became heavy burdens.

Western Islam, centered in southern and central Spain, had reached its peak political and military power by the 11th century at which time Christian Spain, which had managed to survive in the north, began to gain control over larger and larger areas of the Spanish peninsula, a process that reached its climax at the time of Columbus five hundred years ago. In the meantime, a very remarkable cultural amalgam involving Moslems, Jews and Christians began to take place in Moslem Spain, one which was crucial for the diffusion and future development of science.

One of the special examples of such fusion is provided by the work of Rabbi Maimonides (1135-1204) whose family migrated from his native Cordoba to Cairo during a period of political turbulence among the Moslems in Spain. There he prepared a series of books which interlinked Jewish, Arabic and Greek (particularly Aristotelian) culture that has remained a major landmark.

Northern Europe

The 12th and 13th centuries were buoyant ones in Western Europe. New approaches to agriculture and engineering programs that had been initiated in part in Charlemagne's time three and four centuries earlier were coming to fruition. They led, for example, to improved crop yields. Moreover, this was accompanied by the introduction of new tools and labor saving devices which advanced the quality of life. Many new common-place working devices such as screws and drills, as well as windmills and water wheels, were either developed *ab initio* or acquired as a result of observations in Asia. Among other things, the crusades, which had started in 1095 A.D., provided the Europeans with access to new technological concepts which could be adopted to western use with substantial rewards.

The openness of this period in Europe is illustrated by the case of John of Salisbury who in the middle of the 12th century spent twelve years studying at educational centers throughout the Continent in an environment that was congenial and rich in opportunity.

It is remarkable how rapidly the pace and sophistication of scholarship and the practical arts advanced once the wave of transferred Islamic culture began sweeping across Europe. By this time the regions north of the Mediterranean clearly were ripe for receiving new knowledge and fusing it with their own.

In the meantime, in order to advance the interests of the church, and to an extent the needs for professional education, the Popes had authorized the creation of a few universities and study centers for scholars to supplement the well established Italian universities such as Bologna, Padua and Salerno. Initially the University of Paris was the most prominent of these, being regarded by the church as the international center for ecclesiastic studies, although other centers such as Oxford were eventually to gain great prominence. Oxford incidentally grew from a quasi-independent *studium generale* formed by a small band of scholars in the 12th century to a great center in the 13th century. Almost from the start it was sympathetic to scholars interested in science. The growth of the university was accelerated by a ban instituted by King Edward I of England on study at the University of Paris by English scholars because of an altercation with the French.

Step 4—The Transfer

Introduction of Greek and Arab Science

One of the most remarkable incidents affecting the future of science which took place at this time was the arrival in Northern Europe from Moslem Spain in the 12th and 13th centuries of some of the great Greek texts. More specifically, the leading ecclesiastical scholar and authority of his day, Albertus Magnus (1206-1280), became fascinated with the works of Aristotle, particularly those dealing with science. Albertus wrote a number of commentaries on Greek and Arab science. At a conclave in Paris he succeeded in having Aristotle's works integrated into the list of texts acceptable for ecclesiastical studies and critical analysis. He was aided in this process by his former student, Thomas Aquinas, who became a great authority in his own right. Here indeed was an unsuspected Trojan Horse inserted into the official curriculum of Christendom.

It is interesting to note that Aristotle's work held such a prominent place in six great cultures (Greek, Egyptian, Roman, Jewish, Arabic and Christian) for over two thousand years. Indeed, the author has been informed that Aristotle's physics continued to be central to the teaching of that subject at the University of Copenhagen well into the 18th century. This is probably not a unique example.

Bertrand Russell expressed what might be termed indignation at the nature of this endurance, almost as if Aristotle had plotted to dominate all branches of logic and philosophy for all time. Two major factors contributed to the endurance. First, Aristotle had one of the most brilliant minds that has emerged in the Western World and communicated his thoughts in a very scholarly way. It took the work of equally great individuals such as Buridan and Galileo to begin tipping the scales in another direction in a major field of science. Second, as one of the great logicians, some say the inventor of logic, he presented his greatest works, including those on science, in the form of logical essays which could only have been the school master's delight throughout the ages. It is not surprising that his essays had the effect of overwhelming young minds and leaving an enduring impression of infallibility.

It may be added that Albertus Magnus, who had received his initial religious training in Cologne in the Rhineland before advancing by stages to a very major position in the world of church scholarship, was, although unusually interested in science, in many respects not atypical of the clergy of his day in the sense that he had a good appreciation of practical affairs and an admiration for what could be called experimentation. It will be recalled that many monasteries were at least partially self-sufficient in respect to agriculture, manufacturing and building maintenance so that the clergy did not feel far removed from practical manual work. In addition, Albert belonged to that class of religious scholar who sought revelation through the study of sacred and ancient texts as an alternative to what might be termed direct mystical experience.

Apparently, we must conclude that the previous pattern of church scholarship must have become somewhat stultifying to the more brilliantly imaginative scholars of the time. To them, access to the Greek and Arabic work, particularly the scientific work, provided an exciting new challenge in the form of material which could be subject to critical examination using the remarkable talents for logical analysis that had been developed among the scholars in the Church over the centuries. Moreover, they approached the new areas of study with something equivalent to youthful zest in spite of the relatively high level of sophistication ultimately achieved in their traditional scholarship.

In any event, the ancient texts and the Arab additions immediately drew the attention of other scholars such as Robert Grosseteste (1175-1253), Roger Bacon (1214-1294), Thomas Aquinas (1225-1264), mentioned above, and William of Moerbeke (1215-1286). For example, Roger Bacon, a brilliant but eccentric figure, extolled the virtues of experiment and the application of mathematics in science. Much effort went into the process of locating original versions of the Greek work and translating them into Latin. William of Moerbeke, who served as bishop in Corinth was responsible for many of the official translations. Although some of the original Greek Texts had been translated into Latin earlier in church history, they had not fallen on the same fertile ground. Moreover, many of the Arab versions of Greek texts used in Spain had gone through successive translations and suffered distortions.

Had the principal commanding figures of the church in Rome appreciated what lay ahead, they probably would have attempted to ban much of the Greek and Arabic work as heretical as had been the case in earlier centuries. It is not as if heresies had disappeared from Christianity. In fact, the 13th century saw the rise in southern France of the Cathars, a sect with a mixture of retrogressive, puritanical and otherwise disruptive views which were strongly influenced by importations from the Near East as well as Spain. This movement, sometimes termed that of the Albigens with which it was linked, was an off-shoot of Manichaeism, one of the great religious movements of the period originating in Persia. It had adherent sects in areas extending from the Pacific to the Atlantic oceans. It was ruthlessly cut down in a most brutal way with the use of the armies of northern France.

It was in connection with the suppression of the Cathars that the rudimentary structure of what eventually became the Spanish Inquisition was created. The Inquisition incidentally was not disbanded until the 1830's following the execution in Spain of an otherwise harmless school teacher who, although deeply religious, did not handle religious formalities in the classroom in a way that conformed precisely to proscribed rules of the Spanish Church. This incident aroused much resentment in Western Europe.

Buridan: The 14th Century

> We have heard of late that in the lands of the Franks, that is, in the country of Rome and its dependencies on the northern shore of the Mediterranean Sea, the philosophic sciences are thriving, their works reviving, their sessions of study increasing, their assemblies comprehensive, their exponents numerous, and their students abundant. But God knows best what goes on in those parts. "God creates what He wishes and chooses."
>
> Ibn Khaldun (1331-1406) courtesy of Professor Bernard Lewis

In a convocation at the University of Paris, Albert had called attention to Aristotle's claim that, in effect, even God could not cause an arrow to fly in a vacuum after leaving the bow. Since this aspect of the Greek work seemed heretical, Albert recommended that it be given critical reviews and subject to experiment if necessary.

The issue of the arrow's flight was subsequently investigated in great detail in the next century by one of the most brilliant, ordained, ecclesiastical scholars of that century, namely, the Piccard, Jean Buridan (1295-?). He was also one of the most colorful individuals of his time who was reputed to be involved in a broad spectrum of activities, some genuine and some apocryphal, which have become part of historical lore. Buridan noted that when a ship is drifting in otherwise calm air and water, straw loosened from the hand blows to the stern and not the bow. The actual motion of air relative to the ship was not supportive of continued motion of the ship as it would be if Aristotle's surmise was correct. In addition, he set a grindstone in rotation and found no evidence that the air currents around it were of a magnitude sufficient to keep it rotating in the manner proposed by Aristotle after he stopped cranking. He concluded that Aristotle was seriously in error. Here we see a dramatic display of a direct approach to the process of obtaining information through hands-on experimentation, a quality that was so lacking in the greater part of the Greek scientific tradition.

After much cogitation, Buridan decided that the forces which set an object in motion also impart to it a transient agent or property which he termed *impetus* that is gradually dissipated and finally permits the object to assume its natural state, namely, that of rest. The concept of impetus as used by Buridan is not to be confused with the modern concept of inertia. It was viewed in much the same way as the transient warmth that an object at a low temperature acquires if held in the hand for a while and which is gradually dissipated when it is left again at the original low temperature. Buridan may well have picked up the term impetus from readings of St. Augustine or other early work but used it in his own way, not deviating greatly from Aristotle.

Two centuries later the young Galileo, who still had his major creative career ahead, spoke of impetus as being analogous to the dwindling reverberations of a bell after it has been struck. He was clearly influenced by the work of Buridan.

A contemporary German cleric, Dietrich of Freiberg (apparently died in 1311), provided a partial solution to one of the age-old mysteries, the origin of the rainbow. The explanation of this spectacular phenomenon had puzzled many generations and even individuals such as Avicenna and Alhazen admitted ignorance. Dietrich, who was manually skilled, constructed a number of glass spheres filled with water and demonstrated clearly that the rainbow was undoubtedly caused by internal reflection and refraction in spherical raindrops. Significant advances had to await the time of Newton and Descartes several centuries later.

In this same general period, Nicole Oresme (c. 1320-1382) revived the notion that it was as reasonable to believe that the stars were fixed and the earth rotated as that the earth was fixed. He emphasized that a finite atmosphere would be able to move with the earth.

The Age Of Exploration

The 15th century, while perhaps most notable for the beginning of the great explorations such as those in which Vasco da Gama rounded the southern tip of Africa and Columbus discovered America, was also distinguished by the advances in science and technology associated with the stimulation of the new activities.

Typical of the spirit of the age in science were the experiments of Nicholas of Cusa (1401-1464) who among other things carried out continuous measurements of the weight of growing plants and concluded, after making adjustments for the additions of water and plant food, that some of the factors contributing to growth must be derived from the air. Anticipating Copernicus and echoing Heracleides and Aristarchus, he also proposed that the diurnal and seasonal variations in stellar changes were actually in major part a result of the motion of the earth. A new spirit of action and speculation was on the way.

In another direction, Prince Henry of Portugal (1394-1460), rightly called the Navigator, established a scientific institute of navigation at Sagres at the southern tip of Portugal as mentioned in Chapter 2. There he assembled an international group of experts and, using every means of persuasion possible, directed the ships to explore ever further down the West coast of Africa. Earlier Henry had been involved in a military expedition in which Portugal had captured a portion of the African coast in what is now Morocco from the Arabs and was inspired to envision the possibility of a vast extension of the European Christian world.

One of the tragedies of history is that we know far too little about what went on in Henry's institute at Sagres, partly because of the secrecy maintained. The institute was needlessly burned to the ground by Sir Francis Drake's troops in a skirmish during the war between England and Spain in Elizabethan times. Whatever records were maintained in Lisbon were presumably destroyed in the great earthquake of 1755. Fortunately some ship logs are available and reveal that the captains were aware in a very practical sense of the fact that the earth is a sphere and that the equator cuts across central Africa. As was mentioned in Chapter 2, the scholars in Sagres understood much. Doubtless a great deal of knowledge was transferred from Moslem scholars since the Arabs had explored and established colonies along the east coast of Africa.

The 15th century also saw the invention of printing. This permitted a far easier and freer distribution of information in all fields, not least in science. One of the remarkable features of this development is the rapidity with which the process of printing reached a high degree of perfection. A Gutenberg Bible is a work of art. This shows that the underlying bases for the technology of printing, such as metallurgy, paper making, and mechanical construction, were highly advanced by the end of the century. An indication of the state of technology is illustrated in the writings of Georg Bauer (1494-1555), who wrote under the Latin equivalent of his surname, Agricola.

The Rise of the Middle Class

The general enhancement of trade that followed closely upon the heels of the great sea voyages, as well as the wide distribution of information of all kinds made possible by the printing press, accelerated mercantilism and led to the growth of private wealth. Along with this went a great expansion of private enterprise and the growth of the middle class, a growth which was to continue into our own period and which gave greater economic and political freedom to an ever larger component of society. The advance of science was ultimately to benefit from this growth because larger and larger circles of the citizenry came to appreciate the value of the knowledge disclosed by science as well as the importance of the areas of technology associated with it. This ultimately gave science the final push into widespread acceptance as a major component of European culture.

The Rise of Nationalism

Another distinguishing feature of the 15th century is the rise of national feeling, that is nationalism. It appears that the spirit of national identity was first felt in an explicit sense during the Hundred Years War between France and England which began in the first half of the 14th century.

Along with nationalism arose the spirit of national pride which was eventually to play its own role in the promotion of scientific research, first through the recognition and support gained from royalty, but ultimately through broader national support.

Step 5—The Reformation

The Reformation and Beyond

The next two centuries, from 1500 to 1700, were turbulent years for the advance of science and were indeed the ones in which its status as well as its advance were finally assured. The rapid spread of the Reformation proposed by Martin Luther in 1522, which gained the support of the powerful nobles in large segments of Northern Europe, caused great panic in the church and led it to become militant as well as highly reactionary.

The most militant counterforce to the Reformation arose in Spain in the Jesuit movement initiated by Ignatius Loyola. It became closely coupled with the Inquisition and led to some eighty years of brutal struggle in Northern Europe. The conflict started in The Netherlands in 1568 when the Spanish Duke of Alva was given the task of suppressing deviant religious beliefs in the prosperous communities there. It ended with the Peace of Westphalia in 1648. That peace not only marked the permanent division of Europe along religious lines, but also the establishment of different attitudes toward science in different parts of Europe, differences which were not overcome until relatively recently. In particular, Protestant Europe, including England, became in the main the enlightened home of science as well as the driving force for the advance of technology and commerce. The Catholic Mediterranean portion of Europe became relatively static and ceased to be among the leading sites of scientific innovation.

Although the Netherlands and much of central Europe were devastated in the religious wars and the portion of Flanders that is now Belgium lost many of its skilled workmen and associated industries, such as cloth-making, to England, the other portion of the Netherlands, modern Holland, which fought and freed itself from Spain, proceeded to become a major sea power as well as an international commercial, banking and scientific center. Spain paid a heavy price for its war on the low countries. It not only lost much of the revenue from a rich province, but expended wealth gained elsewhere to support its army. Spain was much weakened as a result. France became the dominant military power on the Continent.

Catholic France, Austria and Bavaria were ultimately much less affected by the change than the Mediterranean countries. They contributed to and benefitted from many of the same developments as the Protestant nations. For example, the French Academy of Science was founded in 1666 just after the creation of the Royal Society of London. In effect, the influence of the church was no longer either over-riding or homogeneous north of the Alps. Apparently the demands of statecraft and commerce and rising nationalism in the northern regions of Europe encouraged the leaders to ignore to a substantial degree the reactionary pressures that emanated from Spain and Rome.

Actually, some of the early Protestant reformers, including Martin Luther, were skeptical about the revolutionary concepts arising from science. For example, Luther and Melancthon opposed the publication of Copernicus' main work in

Wittenberg in the mid-16th century. This mood was a transient one, however. Presumably the noble and merchant classes were anxious to benefit from the advantages associated with relative freedom from the church and access to new knowledge of all kinds. They determined the actual course of events, encouraging science, technology and trade.

Copernicus

Let us return to science, emphasizing the highlights. The first major event was Copernicus' (1473-1543) proposal of a heliocentric system. While this work represented a return at least in principle to Aristarchus' theory, with which Copernicus was familiar, it was based on the use of much more sophisticated mathematical tools. It may be debated just how revolutionary Copernicus considered himself to be, for his outlook was by no means an entirely modern one. He probably accepted forms of astrology. There is no doubt however that he felt quite strongly that, inasmuch as the sun was the largest object in the planetary system, it was much more logical to have it at the center. Moreover, he was able to show that the Ptolemaic geocentric system based on the use of epicylces could be simplified to a substantial degree, although not completely, with the alternative assumption, but still using epicycles. He hesitated to publish his great work and in fact seemed somewhat apologetic about it, perhaps because of fear of attack from a now indignantly aroused church. Actually, it was published in Nuremberg (1543) close to the time of Copernicus' death on the initiative of a former devoted student, Rheticus (1514-1576), a German Protestant. Unfortunately an editor, probably to mute potential criticism, introduced a preface suggesting that Copernicus placed primary emphasis on the mathematical simplifications associated with the new work rather than the physical implications. Although Copernicus may have begun his great work as a mathematical task in connection with a revision of the calendar desired by the Church, the actual text leaves no doubt about Copernicus' true opinion. This is made even more explicit in a letter to Pope Paul III accompanying a copy of his manuscript. A translation of the letter appears in the book by Adrian Berry listed in the bibliography at the end of this chapter.

Perhaps Copernicus's greatest contribution to the scientific thought of the time was to propose a break in the traditional outlook at a very critical period and thereby open minds to possible alternatives, which is no mean accomplishment.

Incidentally the great Danish observational astronomer Tycho Brahe (1546-1601), who was born three years after Copernicus' death, never accepted Copernicus' proposal although he did accept the older notion going back to the Greeks that Mercury and Venus revolve about the sun.

Steven, Kepler and Galileo

Simon Stevin (1548-1620), Johan Kepler (1571-1630) and Galileo Galilei (1564-1642) arrived on the stage almost at the same time in the second half of the 16th century. With them we enter into the period in which most of the leading figures in European science are from the lay world. The opportunities for study and research had opened up to a wider audience.

Stevin, a Fleming born in Bruges, experimented with systems in static equilibrium under forces acting in a plane and discovered the law of the triangle of forces. A portion of his work also focused on the nature of forces transmitted in fluids and provided a clear exposition of the relationships governing hydrostatic pressure. In addition he was a brilliantly innovative engineer and played an active role in the defense of what is now Holland in the period of the Spanish persecution. Not least he helped in the development of novel fortifications and guided his compatriots in the manipulation of the sluice gates of the dikes in Holland so as to develop sufficient skill to impede or entrap the Spanish army. Stevin could in truth, be said to be one of the first modern scientists-engineers.

Kepler accepted the Copernican Theory as a result of early indoctrination by one of his teachers and began to study the laws governing planetary motion on this basis. He was guided by an excellent command of the mathematics of his day as well as great personal ability in the field. He is credited with the invention of a form of infinitesimal calculus which he developed for his own use. Associated with his work was the belief, often expressed by others, that planetary cycles as a whole must have some innate harmonious relationship, analogous to a musical scale.

He spent a period with Tycho Brahe in Prague after the latter's exile from Denmark where he had had his great observatory. This period encompassed the time of Brahe's death. In consequence, Kepler was able to take away with him from Prague a complete set of Brahe's measurements of planetary motion, by far the best available in the world at that time. Apparently Brahe's relatives objected to what they looked upon as a theft of this work, but we can be grateful for his temerity. From an analysis of these data, and after much study, he determined that the orbits are elliptical in shape with the sun at the focus of the ellipses. He also deduced the law of areas and the relationship between period and mean radius. Moreover, he concluded, in a more qualitative way, not only that the planets are attracted to the sun in a manner analogous to magnetic attraction but also that the planets interact with one another. This led him to propose the existence of a law of universal interaction of matter.

The fact that Kepler was a Lutheran did not prevent him from serving as astrologer in Catholic royal courts even during the Thirty Years War in central Europe (1618-1648).

Galileo stands out as the greatest individual in the transition between the old and the new. His personal history symbolizes in many ways the final struggle associated with the rise of modern science. In a sense, there are three significantly different phases to his professional career. In his youth, essentially as a student, he expressed fairly conventional views concerning the work of Aristotle. He endorsed, as mentioned earlier, the concept of impetus introduced by Buridan as a solution to the problem of the flight of the arrow. He did, however, study the regularities of the motions of the pendulum and became deeply interested in advancing understanding of the most basic problems of mechanics.

His second period, which in the main centered about the eighteen years spent at the University in Padua (1592-1610), was particularly fruitful because he devoted his efforts to improving experimental techniques and observations as well as speculations. Incidentally, the city of Padua had fallen under the suzerainty of Venice early in the 15th century and enjoyed considerable independence from the influence of Rome. Galileo's first appreciation of the laws governing bodies being accelerated under gravity came at this period. In addition, he learned of the invention of the telescope by Dutch instrument makers, constructed one of his own, and made the primary observations of the moon, the planets and the moons of Jupiter as well as of sun spots.

Galileo's third, troubled period followed his move to Rome. It turned out to be a climactic one in more ways than one. In it he finally achieved a clear understanding of the concept of momentum, as distinct from impetus, and thus came to appreciate the principle of relativity of uniform motion in the sense of what we now term classical mechanics. Moreover, he expanded his quantitative investigations of the way in which a mass accelerates when subject to a constant force. It is interesting to note that Galileo was well advanced in years when the two basic principles governing the motion of bodies were finally clearly formulated in his mind.

Beyond all of this, his observations of the planets and sun with his telescope not only convinced him of the Copernican planetary theory but made him appreciate the fact that planetary matter could not be radically different from that on earth. He also decided that the stars must be very remote from the earth since he could not resolve any of their disks in his best telescopes.

He was the first individual who had an opportunity to appreciate on firm experimental grounds something of the immensity of the universe and to catch a glimpse of the true dispersal of matter in it.

Kepler attempted to communicate with Galileo, whom he admired extravagantly, on a regular basis but the latter put him off, probably because of fear of an unpredictable reaction from the church.

When the Inquisition began to attack Galileo for his belief that the earth rotated and revolved about the sun, the Venetian authorities urged him to return to Padua where he would be protected. Presumably he stayed on, first in Rome and eventually in Florence, because he thought his discoveries were so rationally based that he could overcome criticism. He insisted that he was only revealing the handiwork of the Creator. It is indeed possible that the years he spent in house arrest benefitted his scientific productivity by allowing him to rethink the significance of earlier works but this fact cannot condone the actions of the Inquisition which harassed and threatened him a number of times.

Harvey

We must not forget that this is also the period in which William Harvey (1578-1657) revolutionized the theory of the circulation of the blood, following a period of study at Padua. Medical studies there were still under the influence of the stimulating teachings of Vesalius (1514-1564) who had taught medicine at the university and raised both the standards and traditions of the field to new highs partly through the revival of dissection. There is good evidence that Harvey and Galileo became friends at Padua and that the former probably gained familiarity with some of the principles governing the action of mechanical systems as a result of this association.

Harvey's great work was in part anticipated by that of Michael Servetus (1511-1553), a Spaniard, who proposed a model of the main circulatory system through the body. He was burned at the stake by the Calvinists in Geneva for deviant religious beliefs.

Sir Francis Bacon

Much is made in the English speaking world of Sir Francis Bacon's (1561-1626) interest in the rise of science and of the possible influence his book, *The Novum Organum* on the advance of science. His interest probably represented a reflection of the course of developments rather than a driving force behind them since work such as that of Stevin, Kepler and Galileo was well underway when Bacon's book was written. These comments do not detract from its value as an inspired classic associated with a great genius.

Torricelli, Pascal, Descartes and Huygens

The work and views of Stevin, Kepler and Galileo opened up opportunities for a large group of outstanding individuals in the 17th century. Torricelli (1608-1647) and the cleric Pascal (1623-1662) demonstrated that the weight of the overlying atmosphere exerts pressure at the surface of the earth. Pascal observed that the pressure diminishes rapidly as one goes to higher altitudes so that the atmosphere forms a blanket or "skin" of essentially finite thickness.

Descartes (1596-1650), among other things, introduced new mathematical methods into the study of physical problems. To him we owe the wedding between algebra and geometry which we today know as analytic geometry. Moreover, he was probably the first true popularizer of science in the best sense of the word since, through lectures and essays, he inspired many others to become involved in research.

Huygens (1629-1695), the son of a distinguished Dutch diplomat, was also a splendid mathematician and a good friend of many of the great mathematicians of his day. He is probably best known within the general physics profession for developing new ways of looking at the propagation of waves such as sound or light in terms of regeneration of secondary wavelets at wave fronts. However, he was a true descendant of Galileo in the sense that he developed the understanding of centrifugal force in terms of acceleration and went a long way toward perfecting the pendulum clock, one of Galileo's dreams. In his day he was often referred to as the "modern Archimedes."

It is interesting that Huygens, a Calvinist from an intensely Protestant Holland, found great favor in France where he aided Louis XIV in the creation of the French

Academy of Sciences. In contrast Descartes did much of his work in Protestant countries especially Holland. As mentioned earlier the Northern countries were prepared to come to terms with religious differences when important matters were at stake.

There seems to be little doubt that the potential material benefits, economic and technical, associated with the development of science had begun to be appreciated by both royalty and the merchant classes in northern Europe as important contributions to the standards of life. An enhanced sense of national pride and the spirit of free economic enterprise fostered the continued advancement of science.

Newton

We come finally to Newton (1643-1727). Having been schooled in an English speaking country, the author of this book had concluded that Newton was the ultimate source of revelation with respect to mechanics. It came as some surprise in his early association as a graduate student with Eugene Wigner to learn that the Central Europeans, without denigrating the great genius of Newton, distributed relatively more credit to his predecessors and regarded Newton as a superb mathematician and physicist who among other things summarized, refined and extended in a uniquely creative way the discoveries of his predecessors. Galileo had derived the basic laws of motion and it was already known as a result of Huygens developments concerning centrifugal force, that Kepler's third law governing the periods of the planets would follow for circular orbits if the sun exerted an inverse square attractive force on the planets. Newton, as one of the inventors of calculus, was able to demonstrate Kepler's laws for elliptical orbits thereby opening up a whole new dimension of theoretical mechanics which he exploited in a number of ways. As part of this process he extended the law of gravitation and applied the result to lunar motion. Beyond this he demonstrated that white light is composed of a mixture of the colors of the spectrum, a matter which opened, but ultimately settled, much controversy.

The continental mathematical physicists, who tended to stay in close association with one another, regarded Newton with a mixture of awe and puzzlement. One is reminded of the comment made by the brilliant Swiss mathematician, Bernoulli, on reading a supposedly anonymous solution originating with Newton to a mathematical problem, "One recognizes the lion by the print of its paw."

Whatever else may be said, there is no question that Newton's work brought some five centuries of endeavor to a climax and led to what we can unambiguously call the Newtonian Age of Science without undervaluing the seminal work of those who came before him. From Newton's time onward what we term Western Science occupied an unassailable position in human culture. The pathway to the great revelations which lay ahead was clearly open.

Religion and Science

It is worth commenting on the fact that all of the major contributors to science up to the period which includes Newton were deeply religious. In one way or another, they regarded their work as part of the process of revealing the work of the Creator. This, it will be recalled, was part of Galileo's defense when accused of heresy. It is interesting to note, however, that Galileo, unlike many others, seems to have been relatively free of the relics of Medieval beliefs such as those related to astrology and alchemy. Newton, rather strangely, was deeply imbedded in them. In fact his brilliant contributions to the advance of natural science almost give the appearance of being incidental, as far as he was concerned, to a preoccupation with alchemy as well as religious doctrine and its history. As mentioned earlier, he strongly endorsed the views of Arius concerning the Trinity.

Science and Liberal Democracy

Although modern science as a method of investigating nature came to fruition before liberal democracy came on stage approximately two hundred years ago, it has found a fairly congenial and supportive home in the open democracies so far. While the closed highly centralized societies strongly desire to gain the benefits of the applications of science, basic research cannot be said to rest nearly as easily in such social structures. In brief, even at best, the scientists in closed societies are on a short leash and subject very directly to the whims of the political leadership.

In discussion, Alvin Weinberg has emphasized that the vistas opened up to human exploration by the early successes of science undoubtedly had a catalytic effect on the evolution of the age of enlightenment which in turn inspired the development of liberal democracy.

The Sleepwalkers

The popular author, Arthur Koestler, has referred to the group of transition scientists leading up to Newton as "The Sleepwalkers" and has used this as the title for a book dealing with their work, as if they perhaps were guided somewhat blindly by their subconscious rather than conscious minds. Actually this point of view is fairly superficial, granting that Koestler's book is otherwise a fine contribution to the popular science literature. From Buridan to Newton, those who were successful in advancing aspects of scientific research had very specific objectives and used the facilities available to them, often conditioned by limitations on current technology, to the best of their ability. Yes, there were misguided dreamers, incompetents and charlatans along the way which might imply the environment was different then. But we must not forget that we continue to have all of these in today's world, if we recall the current pursuit of such matters as astrology, flying saucers and apricot pit cures for cancer.

Chinese Science[2]

What of the Chinese conundrum? As we have noted, the practical Romans, undoubtedly under the influence of cultural rather than genetic constraints, were quite willing to permit the Greeks to pursue science to mutual advantage while they pursued other areas of endeavor, not least engineering at which they were superb. Clearly the cultural climate in China did not place high priority on areas of scholarship that might have led to modern science even though there apparently is no evidence available that activities related to science were in any sense forcibly suppressed. We do know that when the Chin unified China in the 3rd century B.C. the leadership did burn certain classical books, particularly Confucian material, as it sought to enforce a new type of unity within the newly created empire. That, however, was a very different matter.

One of the most elegant general proofs of the Pythagorean Theorem concerning a right angle triangle was derived by a Chinese, Chen Kung, in the 2nd century A.D. It might well have been inspired by access to Euclid at that time. A number of Chinese monks and scholars became interested in Indian religion and culture during the Han period which was more or less coincidental with the Roman era. It would seem much more likely that the Chinese interest in such general geometrical proofs developed as a result of relatively direct knowledge of Euclidian geometry through India which had a strong mathematical culture rather than as a result of more indirect contact with the Greeks through the western branch of the Silk Road.

What is obviously true is that the Chinese were extraordinarily gifted at what Alvin Weinberg has termed the technical fix. They made remarkably good use of materials available to them in the various complex situations they encountered, using levels of imagination equivalent to that exhibited by any ingenious practical experimenter. These gifts extended to the field of what one may call either chemistry or alchemy. With this great capability, they seemed to feel little need to reach further in their approach to the physical world even though they could be (and are) inspired by exotic poetry and other matters of the spirit.

What is perhaps most surprising is that their great talent for devising ingenious devices did not lead them to the steam engine. As a concept, such an engine lies well within the range of matters that did interest them.

Chapter 2 of this book is devoted to a discussion of the outer limits which modern technology might have been expected to reach if science had not developed. We need

---

[2] During visits to Taiwan, the author had the privilege of discussing these issues with two eminent scientists there, namely, Dr. K. T. Li, who played a leading role in the technical and economic development of the island and Dr. T. Y. Wu, the president of the Academia Sinica. Dr. Li commented wryly that Chinese tradition was such that an individual who discovered a scientific law would be inclined to keep it a family secret and it would probably get lost in the various transfers. Dr. Wu more philosophically pointed out that Chinese culture has always been highly pragmatic and basically interested in day-to-day affairs. Confucianism, the traditional basis of proper conduct, is designed to maintain continuity in society and the development of conflict-free human relationships. The pursuit of science is not only traumatically evolutionary but destabilizing and hence not well in phase with the Confucian code of ethics.

not have been denied the age of steam power, including railroads and the like. James Watt was an engineer not a scientist. He did not rely upon science for his inspirations. Apparently, we seem forced to conclude that the failure of the Chinese to take full advantage of the opportunities inherent in science and technology rests not on the lack of creative minds of great genius with the requisite technical abilities, but upon a constriction in the range of accepted values on the part of individuals, determined by special features of the cultural climate in the great Middle Kingdom.

It has sometimes been said that the existence of slavery inhibited older civilizations from taking an interest in either more advanced technology or science. One can seriously question this suggestion since more advanced technology permits one to accomplish tasks and goals that would otherwise be impossible. Moreover, most societies have had a continuing interest in concepts and instruments that might prove advantageous in warfare. It should be added that Greece, Rome and Islam were slave societies during the formative stages of science.

We must also remember that individual curiosity, often working without practical ends in mind, has always been a driving force for innovation. Indeed it seems safe to say that the most basic forms of science did not really yield true material profits until the 19th century when fields such as chemistry, optics and electromagnetism began to generate large industries.

It has also been said that the evolution of monotheism provided an incentive for the development of understanding of possible universal principles. Here again one can question the thesis. The Greeks were not monotheists, moreover it must have been evident to the Chinese in spite of their complex pattern of religious beliefs that the ingenious devices they invented behaved in essentially the same way in one place as well as another. Engineering principles have a high degree of universality, not unlike those of science.

If there were miracles on the pathway to the development of modern science, beyond the interest exhibited by the Greeks and Arabs, there were two principal ones. First, the reactionary church leaders were much too late in recognizing the dangers associated with accepting the Greek-Arab work in the fields of science to stop its further evolution in the Christian world where it has thrived. Second, the Western Europeans possessed two qualities found separately in the Greeks and the Chinese but which were not common to both, namely, a profound interest in natural phenomena, viewed on an intensely logical basis, and a willingness to devise specially designed equipment that might be used for experimental work in pursuit of further and more complete knowledge.

Bibliography

G. Abetti, *The History of Astronomy* (Abelard-Schuman, New York, 1952).

A. Berry, *Scientific Anecdotes* (Harrap, London, 1989), Chapter 3 (page 62) contains a translation of Copernicus' letter to Pope Paul III.

D. M. Bose, et al., Eds., *A Concise History of Science in India* (Indian National Science Academy, New Dehli, 1971).

E. Bradford, *Southward the Caravels* (Hutchinson, London, 1961).

E. Brehier, *The Hellenistic and Roman Age* (University of Chicago Press, Chicago, 1965).

C. B. Boyer, *The Rainbow* (Princeton University Press, New Jersey, 1987).

H. Butterfield, *The Origins of Modern Science* (Macmillan Co., New York, 1962).

J. L. E. Dreyer, *A History of Astronomy* (Dover Edition, 1953).

P. M. M. Duhem, *Le Systeme du Monde* (Hermann, Paris, 1913-1959).

W. Eberhard, *A History of China* (Taiwan Press, 1971).

L. Fermi and G. Bernardini, *Galileo and the Scientific Revolution* (Fawcett Publications, Greenwich, Conn., 1965).

O. Gingerich and R.S. Westman, *The Wittich Connection: Conflict and Priority in Late Sixteenth-Century Cosomology* (Volume 78, Part 7, Transactions of the American Philosophical Society, 1988).

T. Heath, *Aristarchus of Samos* (Dover Publications, New York, 1981).

F. Heer, *The Medieval World* (World Publishing Co., Cleveland and New York, 1962).

A. Koestler, *The Sleepwalkers* (Macmillan, New York, 1959)

I. Lapidus, *A History of Islamic Societies* (Cambridge University Press, Cambridge, 1988).

J. A. Lekstrom, *Medical Literature of Medieval Salerno* (Pharos, Winter 1990), p. 91.

B. Lewis, *The Muslim Discovery of Europe* (W.W. Norton and Company, 1982).

A. A. Maurer, *Medieval Philosophy* (Pontifical Institute of Medieval Studies, Toronto, Ontario, Canada, 1982).

W. H. Mitchell, *The Rise of the West* (University of Chicago Press, Chicago, 1963).

J. L. Motley, *Rise of the Dutch Republic* (Reprinted, three volumes; The Temple Press, Letchworth, 1950).

S. Nakayama and N. Sivin, *Chinese Science* (MIT Press, Cambridge, 1973).

O. Neugebauer, *Astronomy and History*, Selected Essays, (Springer-Verlag, New York, 1983).

O. Neugebauer, *The Exact Sciences in Antiquity* (Brown University Press, Rhode Island, 1957).

G. Parker, *The Army of Flanders and the Spanish Road* (Cambridge University Press, Cambridge, 1987).

S. Pines, *Saint Augustine et la Theorie de L'Impetus*, Collected Works (The Magnes Press, The Hebrew University, Jerusalem, 1986), Vol. II, p. 394.

J. Plaidy, *The Spanish Inquisition* (Citadel Press, New York, 1967).

H. T. Pledge, *Science Since 1500* (His Majesty's Stationery Office, London, 1939).

A. I. Sabra, *The Andalusian Revolt Against Ptolemaic Astronomy*, Transformation and Tradition in the Sciences, Everett Mendelsohn, Ed. (Cambridge Press, 1980), p. 133.

A. I. Sabra, *The Appropriation and Subsequent Naturalization of Greek Science in Medieval Islam* (Science History Publications, Ltd., London, 1987) Hist. Science, Vol. XXV, p. 223.

A. I. Sabra, *Avicenna on the Subject Matter of Logic* (The Journal of Philosophy, 1980), p. 746.

A. I. Sabra, *Islamic Optics*, Dictionary of the Middle Ages (Charles Scribner's Sons, New York, 1987), Vol. 9, p. 240.

A. I. Sabra, *Islamic Science*, Dictionary of the Middle Ages (Charles Scribner's Sons, New York, 1988), Vol. 11, P. 81.

A. I. Sabra, *The Scientific Enterprise*, Bernard Lewis, Ed., The World of Islam (Thames and Hudson, London), Chapter 7.

A. I. Sabra, *Psychology Versus Mathematics: Ptolemy and Alhazenon the Moon Illusion*, Mathematics and Its Applications to Science and Natural Philosophy, E. Grant and J.E. Murdoch, Eds. (Cambridge University Press, 1987), pp. 217-247.

S. D. Sargent, *On The Threshold of Exact Science—Selected Writings of Annaliese Maier* (University of Pennsylvania Press, Philadelphia, 1982).

N. Schachner, *The Medieval Universities* (A.S. Barnes and Co., New York, 1962).

F. Seitz, *Science Futures*, Science and Technology in the World of the Future, A. B. Bronwell, Ed. (John Wiley and Sons, New York, 1970).

C. Singer, *A Short History of Scientific Ideas* (Oxford Press, 1959).

R. Taton, *Ancient and Modern Science* (Basic Books, New York, 1963).

R. S. Westfall, *Never at Rest—A Biography of Isaac Newton* (Cambridge University Press, Cambridge, 1980).

J. G. Yoder, *Unrolling Time—Christiaan Huygens and the Mathematization of Nature* (Cambridge Press, 1988).

4
The Divine Fire Crosses the Atlantic Ocean[1]

There is a story to the effect that a mutual friend introduced the young Franz Schubert to the aging and deaf Beethoven. The former showed him some sheets of music. Beethoven perused them carefully and then said, "Young man, you possess the Divine Fire." The Divine Fire—what a glorious term! According to Greek mythology, Prometheus stole it from the gods and gave it to mankind to provide it knowledge and wisdom, an act for which he was severely punished. Of all the forms which the Divine Fire can take, and the great forms of artistic achievement are one, there has been none more important than the gift mankind has received through the development of basic science. Today, the torch that represents that aspect of the Divine Fire burns brightly in North America. How did the transition across the Atlantic ocean take place? Did it come with the initial colonists or was it a much slower and more laborious process? The facts make it clear the second route was followed. The divine fire did not glow brightly at first. Many dedicated scientists had to struggle before recognition finally came.

1. The Start

When pre-Revolutionary antecedents decided to settle North America, they faced a very formidable problem because the land was a wilderness which had to be tamed by whatever means the labor and ingenuity of man could devise. It followed that

---

[1] This chapter, strongly influenced by the author's Washington years, does not do justice to the development of science in neighboring countries in North America, particularly Canada. As is evident from the discussion in Chapter 9, McGill University in Canada, founded in 1821, was an important center for science in the 19th century. In fact, nuclear physics had its birth there as a result of the work of Ernest Rutherford and Frederick Soddy at the turn of the century. The University of Toronto, founded in 1827, has an excellent record of scientific achievement, particularly in the 20th century. For example, the importance of insulin in the treatment of diabetes was discovered there in the 1920's. Science in Canada was stimulated initially by its close links to the United Kingdom.

the colonists and the pioneers, migrating for the most part from the regions of northern Europe where science and engineering were valued highly, placed major emphasis on the applied arts in the culture they brought with them. The more refined aspects of basic science, of which the intellectuals and enlightened aristocracy in Europe were so proud because they represented a source of new vision as well as technology, were kept in the background. It was a land in which development and invention were given far higher standing than research which would advance knowledge for the sake of enlightenment and the possibility of eventual use.

It is true that there were a few special circumstances that should be noted. The young gentleman who received an education at one of the relatively newly formed colleges in the colonies, or later on in the newly independent country, could be expected to take pride in some understanding of science. Emphasis was given to mathematics, probably algebra and geometry, or in some cases the elements of physics and astronomy. Perhaps the supreme example of an individual bearing this form of culture is provided by the gentleman plantation owner, Thomas Jefferson. He prided himself on his broad knowledge of scientific and technical matters and operated both his home and his plantation with much technical ingenuity and inventiveness when not otherwise preoccupied.

2. Benjamin Franklin

There was, of course, one great individual in the colonies who was truly a world class scientist, namely, Benjamin Franklin. He is best known to the American public as a patriot, printer, publisher, postmaster, collector of aphorisms and diplomat but he also possessed a deep and creative interest in the advancing science of his day. It was he who demonstrated that lightning is a form of electrical discharge and developed the two fluid theory of electricity, giving the names to positive and negative charges.

Franklin also carried out an experiment which shows that his scientific insight knew few limitations. He noted that when oil is dropped upon water, it spreads to cover a certain area and then stops spreading depending upon the amount of oil involved. He did experiments on a pond in England and found that the spreading stopped when the layer of oil reached a definite thickness of the magnitude of what today we would call molecular dimensions. The inference was clear to him. The oil was composed of something which had a definite minimum size. This experiment was carried out long before Dalton proposed the atomic theory. For this and his other work, Franklin was elected a member of the Royal Society of London.

3. The Academies; Government Institutions

When the early settlers formed what might be called scholarly academies, they did not give primary emphasis, in a formal way, to basic science. They tended to stress the practical in comparison to the broader goals of their European counterparts.

Whereas the Royal Society of London, founded in 1660, has the goal of "improving natural knowledge," Benjamin Franklin's American Philosophical Society, founded in Philadelphia in 1744, has the goal of "promoting useful knowledge." The charter of New England's American Academy of Arts and Sciences (1780) had a similar practical motive.

This intensely practical trend shows up repeatedly in the early days of nationhood. The few institutions of a federal character formed prior to the Civil War that might be regarded as scientifically oriented had obvious practical goals. There was, for example, the Public Health Service (1798) which was initially concerned with the health of merchant seamen who could arrive in the ports sick. There was a Coast Survey (1807) which had the goal of charting coastal waters for purposes of navigation. Actually, it served for a number of years as a useful base or haven for much good science.

In the period prior to the Civil War (1861-1865) a number of notable attempts emerged from American soil to couple the strength of the government to the support of basic science. However, all but one of these failed for two reasons. First, a tough-minded Congress, composed to no small extent of men from rural and frontier areas, was skeptical of federal expenditures for scholarly work removed from their everyday interests. Second, sectional fears of the growing strength of the Union caused many representatives to oppose the creation of national institutions. One of the most notable failures was the National Institute which was promoted about 1840 by a group in Washington to provide our government with an active center for work in science and related areas. Although started with much enthusiasm and a smattering of private money, it ultimately died for lack of federal support or even of recognition.

4. The Smithsonian Institution

The one remarkable exception in the entire pre-Civil War history is the Smithsonian Institution which was founded as a government institution in 1846 and had the very distinguished physicist, Joseph Henry, previously of Princeton University and a scientist of international stature, as its first director, or as the position was called then and since, its first secretary. In this case, however, the money, about a half million dollars in gold, which launched the Smithsonian was the gift of an English scientific chemist, James Smithson, who had died in 1829 and left his wealth, inherited from his mother, to the government of the United States to found an institution devoted to science. Smithson, who was the illegitimate and unrecognized son of an English nobleman, had never visited the United States but apparently was inspired by the thought that this relatively classless nation probably had a major destiny in the advance of mankind and that it would inevitably find science an indispensable adjunct to government. Smithson's bequest did not arrive in the United States until 1836. It took Congress a decade to decide whether the money should be accepted and to what actual use it should be put. Fortunately, the men who led Congress ultimately accepted the money and spent it wisely—after sufficient debate. The story is a complex one for there were, as would be the case now, many diversionary temptations along the way.

It will be to the eternal credit of the Congress that decided to move ahead with the creation of the Smithsonian Institution that in searching for a director it looked over the best scientists in the country and made a superb choice, namely, Joseph Henry who was then a professor at Princeton University.

It would be difficult to overstate the value of the Smithsonian Institution in the evolution of federally supported science in the country since it provided a haven for several generations of inspired leaders who played an enormous role in the development of new institutions.

The sectional objections in Congress to the establishment of the Smithsonian as a federal organization were circumvented to a significant degree by emphasizing the concept that the new institution could be viewed as an activity of the District of Columbia, that special square of land cut out of Virginia and Maryland in which most of the governmental buildings resided and which was regarded as common property to all the states. This technicality made it more palatable to those who were concerned about infringement upon the rights of the states.

5. Joseph Henry

Joseph Henry, the first head of the Smithsonian, deserves special mention at this point since he was a very remarkable individual and the greatest scientist in United States after Benjamin Franklin. He was born in 1798 in Albany, New York and attended the Albany Academy where he also became a member of the faculty. His interests, which were initially very varied, eventually focussed on electrical phenomena. During this period he discovered magnetic induction—the generation of an electric current in a circuit by a changing magnetic field—several years before it was discovered independently by Michael Faraday. Henry's priority is recognized today by the fact that the unit of magnetic induction bears his name. He also invented the transformer which is widely used for stepping alternating current voltages up or down.

In 1832 he received an appointment as professor in the field of natural philosophy (termed physics at present) at what is now Princeton University. He continued lecturing and carrying on research there until 1846 when he was asked both to organize and head the Smithsonian, an appointment he held for the next thirty years. He not only made that institution into an excellent research institute but he became the central figure in Washington for promoting the advancement of basic science in the country as a whole. He became a close and much admired friend of Abraham Lincoln during the Civil War.

6. The Civil War

The outbreak of the Civil War in 1861 not only spurred the states which remained in the Union to greater interest in matters related to science and technology but also removed from Congress those who were most bitterly opposed to the establishment of federal institutions. Thus the Civil War years were marked by two very significant actions by the federal government which were to cast long shadows ahead in the evolution of science in the country. The first was the passage of the Morrill Act in 1862 establishing the land grant universities in the states which wished to have them. The second was the creation of the National Academy of Sciences in 1863 as a body intended both to promote science in the nation and to advise the government upon request in areas related to "science or art"—the applied arts being implied. The Morrill Act was, of course, a stupendous instrument since it lent incalculable practical support to programs destined to make higher education available in the long run to all who merited it. Its value can be appreciated by the realization that the country is now entering into an era in which the vast majority of those receiving higher education will pass through institutions which benefitted in a major way from the Land Grant Act.

7. The National Academy of Sciences

The National Academy of Sciences was organized by several of the most dynamic figures of the day involved in political, military and scientific circles. Its Articles of Incorporation were generated by a committee in which Commodore C. H. Davis, who had been active in the naval operations in the Mississippi, took a prominent part. He was aided by outstanding scientists such as Alexander D. Bache, the great grandson of Benjamin Franklin, and Louis Agassiz, a native born Swiss who had joined the faculty of Harvard University. The bill was introduced into Congress by Senator Henry Wilson of Massachusetts who had been an Abolitionist in his earlier days but who, like many of those in his generation who initially fed the sources of sectional strife on both sides of the Mason-Dixon line, became a moderate as he realized the dangers of war. Bache was the first president of the Academy. The initial group of members went to work rapidly on a number of applied science problems of relative urgency in the war.

The organizing group acted with some degree of secrecy in the early stages, excluding Joseph Henry from the circle, presumably because they felt that he would dominate it. Actually, he became a very strong supporter of the organization and eventually its savior when its initial mission, associated with the Civil War, came to an end.

8. Scientists of the Civil War Era

It may be of interest to the reader to have a picture of the background and professional activities of a few of the most prominent scientists who were involved in the formation of the National Academy of Sciences. Generally speaking, whatever their concern in basic science may have been, they usually had some significant association with technological matters in accordance with the spirit of the times. Many of those who had a strong relationship to physics or mathematics were trained as engineers. Many others started out with a medical degree to provide something in the nature of a safeguard in the event that other opportunities did not open up for them. Many of those who were born in United States and acquired engineering degrees did so either in the Army or the Navy which provided a good entre for engineering training. Very few of the scientists of this era focussed on basic science from the start as was the case for Joseph Henry. While there were notable exceptions, such as Henry, most of the outstanding members had spent a significant period studying abroad. All, however, were strongly influenced by and looked up to European science. Many studied in German universities for two reasons. First they were good and welcomed foreign students. Second the research student could obtain a distinguishing doctor's degree with reasonable effort.

Louis Agassiz (1807-1873)

Louis Agassiz was born in Switzerland and displayed at an early age a deep interest in zoology. He undertook higher education at a series of German universities and received there both the Ph.D. and the M.D. degrees, making in the process many warm friends among students and faculty involved in various areas of basic research. At that time, particularly in Switzerland, a deep scientific interest was beginning to develop in the history of the glaciers. They appeared to have varied a great deal over the period of recent human history. Agassiz' first major scientific activity was focussed on this subject and he quickly became one of the leading theorists and expositors in the field. He carried this along with ever-widening concerns regarding many other topics related to natural history. He went to the United States in 1846 as a lecturer and found that he had a significant role to play in the developing country.

He eventually obtained a chair at Harvard where he had a broad opportunity to explore his varied interests. Among other things, he and a son became involved in mining operations in the upper Mid-West and expanded their personal income through this channel. He lectured widely around the United States and had a significant influence on many budding students. One of these was the great naturalist, John Muir, who encountered Agassiz when he was a student at the University of Wisconsin. Muir eventually went to California and became one of the creative exponents of modern glacial theory as a result of observations in the Sierra Nevada Mountains. Amusingly enough, when Muir reached the conclusion that Yosemite Valley was created by glaciers rather than by some catastrophic collapse of the floor of the valley, he was ostracized by the "leading" geologists of the state for advancing such a radical proposal.

Alexander Bache (1806-1867)

As mentioned earlier, Alexander Bache was the great grandson of Benjamin Franklin. He entered West Point at the early age of fifteen to study engineering and was so brilliant that he received an appointment on the faculty after graduation.

This was followed in turn by an appointment at the University of Pennsylvania where he was chosen to give the principal lectures in both physics and chemistry. In the interlude he spent considerable time in Europe both attending lectures and making friends. Ultimately he became head of the United States Coast Survey and extended the scientific work of that organization in a most impressive way. It was Bache, one of the most influential individuals with a scientific background in the Washington area, who finally persuaded Joseph Henry to leave Princeton and organize the Smithsonian. Subsequently he, Agassiz, and Commodore Davis played the leading role in the creation of the National Academy of Sciences, as mentioned earlier.

James D. Dana (1813-1895)

Dana, who grew up in upstate New York, became attracted to natural science and entered Yale University where he focussed on geology under the influence of Benjamin Silliman (1799-1864), another founding member of the Academy. He soon found that he was greatly intrigued by the field of mineralogy and, as a result of investigations in this discipline carried on on a world-wide basis, he became a prominent international authority. In the course of his studies, he prepared reference books on the field which became primary sources for many generations of mineralogists. His work was sufficiently outstanding that he was awarded the Wollaston Medal of the Geological Society of London and the Copley Medal of the Royal Society of London.

Commodore Charles H. Davis

Admiral Davis received his basic education at Harvard College but then entered the Navy in which he was involved in a number of exploratory missions, particularly in the Pacific Ocean. In the process he acquired considerable practical knowledge of the fields of astronomy and hydrography. He then took a shore-based assignment for the Navy in the Coast Survey where he made significant scientific contributions, particularly in matters connected with the influence of tides on the deposition of sand. Eventually he returned to seagoing service and was very active in Naval operations in the Mississippi River during the Civil War. The seagoing activity was interrupted in 1863 by an appointment as Chief of the Bureau of Navigation in Washington where, as we have seen, he became involved in the formation of the Academy.

Wolcott Gibbs (1822-1908)

Gibbs was a New Yorker who entered Columbia University and carried his education through medical school. Having interests in Chemistry, he spent several years in Europe in postdoctoral research and eventually focussed on analytical and inorganic chemistry. On returning, he received an appointment as professor at what is now the City University of New York (founded in 1847) where he was involved in teaching and research in the field of Chemistry. In 1863, he received a professorship at Harvard College both to teach chemistry and as director of the chemical laboratory at the scientific school associated with Harvard. When this school was discontinued, his instruction was limited to post-graduates with associated use of a small laboratory that he established on the banks of the Charles River. He served as the fifth president of the National Academy of Sciences. More about his life will be mentioned in Section 11.

Benjamin A. Gould (1824-1896)

Benjamin Gould became interested in mathematics and astronomy while at Harvard University and decided to continue his education in Europe. Among other things, he was the last student at Goettingen of the great mathematician Carl F. Gauss. His brilliance was sufficiently marked that he was offered the chair in astronomy at the university. However, he preferred to return to United States with the hope of obtaining a suitable appointment there. His first activity involved research with the Coastal Survey in which he helped establish the longitude of a number of places in the country. This was followed by an appointment in astronomy at Harvard. However, he left it in 1870 to move to Argentina where he spent fifteen years as head of the Astronomical Observatory at Cordoba. On returning, he brought with him a large number of photographs of the southern skies which contained extensive material he then published in the Astronomical Journal. For this work he received many honors both at home and abroad.

Arnold Guyot (1807-1884)

Arnold Guyot was also born in Switzerland. Like Agassiz, he developed an early interest in the natural sciences and also went to Germany for higher education. Here he had the good fortune to meet Agassiz as a fellow student. After a period in France, he received an appointment to a chair of history and physical geography at a special postgraduate school in Switzerland. Unfortunately, the Social Revolution of 1848 changed the situation dramatically and the school was closed. At this point, Agassiz recommended that he come to United States. He began there as a lecturer but eventually received an appointment as professor of physical geography and geology at Princeton. He became a close friend of Joseph Henry, who was then at the Smithsonian, and developed close ties with that institution. In addition to creating a significant museum of geology at Princeton, he worked very closely with the Smithsonian in developing special scientific instruments with particular emphasis on the barometer. He used high precision barometers to determine the elevations of a number of mountains on the continent.

Benjamin Pierce (1809-1880)

Pierce, a New Englander, studied at Harvard. While there, he fell under the influence of Dr. Nathaniel Bowditch (1773-1838) who was creatively involved in many fields of physical and mathematical science and was the author of renowned books on navigation. He subsequently received an appointment at Harvard, first as professor of mathematics and physics to which the chair in astronomy was eventually added. He wrote many valuable handbooks and textbooks in these fields that were widely used by students throughout the country. In the course of this work he became a close friend of Alexander Bache and was eventually appointed head of the U.S. Coastal Survey, a post which gave him a great deal more freedom for research.

William B. Rogers (1804-1882)

William Rogers grew up in Philadelphia and received his education at William and Mary College in Virginia where he was ultimately appointed professor of chemistry and physics. This was followed by an appointment at the University of Virginia as professor of physics and geologist for the State. His brother, Henry, was a working geologist and the two made valuable contributions to the knowledge of geological features of the Appalachian chain.

During this period, Rogers developed a plan to create a leading technical school in the State of Virginia. Unfortunately, much destabilizing unrest developed among the students in the southern institutions in the 1850's and he had to abandon the original concept. His wife, a New Englander, persuaded him that he might have a better opportunity for success in Massachusetts. As a result, he moved to that state where he received a number of appointments including that of advisor to the State government on a variety of issues including the study of standards for gas meters and the composition of the gas then commonly used for street lamps and occasionally in households.

Finally, in 1858 he established what is now the Massachusetts Institute of Technology which carried the original title of The Boston Institute of Technology. He became its first president and held that post until his death. Amusingly enough, he warned against ever moving the institution to Cambridge where it would encounter unruly Harvard students. He was also the third president of the National Academy of Sciences, following the death of Joseph Henry.

9. Post-Civil War

Immediately following the Civil War, the Academy tended to flounder in search of new purpose. The uncertainties of this era were compounded by the illness of Bache. In 1868, however, Joseph Henry, then seventy years old and still very active as head of the Smithsonian, became the second president. He decided that the Academy would achieve its purpose in the service of science and the public interest most effectively at that time if it devoted its activities to the support of the highest standards of scientific scholarship which the country could put forth. Henry not only felt that the nation needed to emphasize such standards if it was to achieve levels of accomplishment in basic science comparable to the best standards of

Western Europe, but also felt that the interests of the government would, on the whole, be served most effectively if those who advised it had those qualities of impartiality and character needed for productivity and good scientific work. Among other legacies which Henry left to the Academy during his tenure as its president (1868-1878), was a close association with the Smithsonian Institution. The Smithsonian was the home of the Academy until 1924 when it moved into its present quarters at 21st Street and Constitution Avenue, a building for which a gift from the Carnegie Corporation (a private foundation established by the philanthropic industrialist, Andrew Carnegie) played a prominent role.

10. Science and the Federal Government Prior to World War I

Following the Civil War, the interest of the federal government in science advanced gradually but with ever increasing sophistication as the demands placed upon its technological base increased steadily. During this period, for example, the Coast Survey, renamed the Coast and Geodetic Survey, came to its maturity (1871-1878). The Geological Survey was created in 1879 to advance the exploration of the lands west of the Mississippi. The Weather Bureau was established on the foundations of the earlier Weather Service (1891) and finally the National Bureau of Standards (recently renamed the National Institute of Standards and Technology) was founded just after the turn of the century (1901). The roots of all such organizations can be traced quite far back in early history to institutions created to satisfy very special needs of immediate practical importance. As might be expected, the committees of the National Academy of Sciences played an important guiding role in each of these steps of creation, helping to establish new goals and to resolve some of the conflicts between governmental organizations as new seats of responsibility within the government were being determined.

Of the five presidents of the National Academy of Sciences who served between the death of Joseph Henry in 1878 and the outbreak of World War I in 1914, William B. Rogers (1878-1882) was both a physicist and geologist as well as the founder of the Massachusetts Institute of Technology, O. C. Marsh (1883-1895) and Alexander Agassiz (1901-1907) were field scientists, and Wolcott Gibbs (1895-1900) and Ira Remsen (1907-1913) were chemists. These choices clearly show the attention given in our country during that period to the evolution of technical education, the field sciences and to chemistry. It was, in fact, in this period that the United States began to approach preeminence in fields such as paleozoology and paleogeology and began to lay the foundations for chemical education and chemical research that were to blossom so remarkably and so rapidly after 1914.

O.C. Marsh, the strong-willed director of the Peabody Museum, who discovered, among many other fossil vertebrates, fossils of teethed birds representing critical links between the reptile and the bird families, was an emphatic believer in the view that the Academy should inject opinions into the government. It is said that before going to the Academy offices at the Smithsonian Institution during his regular visits from New Haven in order to handle the affairs of the Academy, he usually stopped off at the White House in order to make certain that both the staff and the President there were well-informed of the views he and other colleagues had generated.

Observational astronomy emerged in a unique way in the United States in the period between the Civil War and World War I. For example, the Lick Observatory, the first mountain top observatory, was established in California in the 1880's. It appears that Benjamin Gould visited California early in that decade and convinced the Bay Area philanthropist, James Lick, of the desirability of financing such a telescope on one of the higher mountains in the nearby coast range. That telescope set the pattern for the Mt. Wilson Observatory (1905) and later on that on Mt. Palomar in the southern part of the state. Similarly the Yerkes Observatory was established in Wisconsin in the 1890's. Bold steps such as these were eventually to place the country in the leading ranks of world astronomy.

Along with the growth of the scientific population came the founding of some major scientific societies. For example, the American Chemical Society was founded in 1874 and the American Physical Society in 1899. The biological community tended to form smaller more specialized organizations as various fields became prominent.

11. Academic Science Prior to World War I

What was happening in the academic institutions during this period? The story is a very mixed one with fairly homely beginnings. Most of the colleges founded prior to the American Revolution were private and tied initially in one way or another to religious denominations. One notable exception was the University of Pennsylvania founded by Benjamin Franklin in 1740 which was intended to be an institution for general learning. In keeping with the demands of the time, the colleges tended to serve four principal functions: preparation for the ministry, law, medicine or general education for gentlemen. Education in science was incidental and limited to a few basics as we have seen. Those who wished to be engineers usually apprenticed to some practical endeavor.

With the founding of the Republic, the opportunities broadened somewhat. As mentioned earlier, the military academies offered some aspects of technical education for those interested in engineering. In addition some of the states established state supported universities—Georgia 1785, Virginia 1791, North Carolina 1795, Michigan 1817 and Wisconsin 1848. Of these, the University of North Carolina was probably the first truly functioning one. In addition to providing general education, these institutions tended to focus substantially on agriculture and engineering in their initial period.

With the growth of a need for technical education, some private technical institutions were started. One of the first of these was Renssalaer Polytechnic Institute founded in Troy, New York, then a major manufacturing center.

To repeat a point made earlier, the representatives of the southern states objected to the federal support of education prior to the Civil War. This restriction was removed by the War, whereupon the Morrill Act granting large tracts of federal land to aid in the establishment of state universities was passed in 1862. This act led to a great expansion of state systems. By the time of the outbreak of World War I, the overall education system had expanded to the stage where approximately ten percent of all secondary school graduates could enter college.

The attitude of the old line private universities with respect to science is interesting. A few, such as Princeton and the University of Pennsylvania, encouraged the development of science departments and associated research while retaining a strong base in what is termed classical education. It will be recalled that Princeton University provided a professorship to Joseph Henry in 1832. From that time onward it proceeded in a reasonably deliberate way to build up the various scientific departments in the university as funds and opportunity permitted. This does not mean that the process was entirely smooth since it was not unusual for incoming presidents at Princeton to decry what they personally considered to be the deleterious effects of science upon our civilization and to express the goal of slowing, or even terminating, the advance of science on the campus. Woodrow Wilson, for example, was among the most outspoken of such presidents. He felt not only that the benefits derived from the advancement of science were dubious but feared what he termed the corrupting influence of research on the field of the classics. He believed that the latter would be tainted by the introduction of scientific methods of investigation. One can only imagine his chagrin if he had lived to witness the situation in which Shakespearean scholars place the language of the great plays into computers in order to check on drifts in vocabulary over time and to compare Shakespeare's style with that of individuals such as Francis Bacon and Ben Jonson who, among others, were, at various times, reputed to be the true authors of the plays. Wilson resigned as president of the university when he failed to thwart the plan of dean of graduate studies (Andrew West) to establish an independent residential graduate school in which it was clear science and scientific research would inevitably play an important role.

The great mathematician John von Neumann, of Hungarian origin, who spent many years at Princeton, first on the Princeton faculty and then at the Institute for Advanced Study, once made the comment that Princeton was proving to be a very safe haven for scientists since science had flourished in spite of the fact that so many incoming presidents had expressed the intention of contracting it.

Attitudes toward science exhibited by the administrations and faculty at some of the older institutions during the last century is indicated by the experiences of the chemist, Wolcott Gibbs, for whom a brief biographical sketch is given in Section 8. Although he was effectively employed by what is now the City University of New York with ample freedom to carry out chemical research, he had a strong desire to be on the faculty at Columbia University. His case was reviewed by the administration of that institution but turned down because it was believed that he would undoubtedly continue to carry on research. Subsequently he obtained an appointment at a scientific school linked to Harvard University as was mentioned earlier. When, however, the university decided to terminate the scientific school, Gibbs was retained on the Harvard faculty but not permitted to teach the undergraduates because he wished to continue research. His research was confined to a small, essentially off-campus shed where he was allowed to work with occasional post-graduate students. Indeed, he carried out some excellent work in spite of the handicaps.

This remarkable attitude of a segment of the humanists in the older private universities in United States toward science and its applications has persisted to a degree even to the present time in spite of the fact that their scientific faculties have, in general, added much luster to the reputations of the institutions on a national and international basis. It is said, for example, that when Yale University was selecting names for its newly established residential colleges in the 1920's, the committee making the decision rejected the name of the great J. Willard Gibbs, one of the most brilliant mathematical physicists the United States produced in the 1800's. The rejection was not made directly on the basis of his profession but because the word "Gibbs" was used as a term for monkeys in old english. Modern day parallels can readily be found.

12. Turning Point

A major turning point of a kind in the status of science in the universities occurred in the last quarter of the 19th century as a result of the development of two new institutions. First was the creation of The Johns Hopkins University in Baltimore with a full fledged graduate school that used the German university system of the period as a model. It was funded by a wealthy Baltimore merchant whose name was given to the institution. While each department had only one professor at that time, it set a new and distinctive standard among private universities.

The second event was even more dramatic. The oil magnate, John D. Rockefeller, Sr., provided the funds necessary to transform a small Baptist institution in Chicago into a major university (1891), namely, the University of Chicago. From the start it had a system of departments that involved more than one professor in each major field. Rockefeller, who had an infallible record for locating excellent advisors, was guided in this enterprise by an unusually gifted individual, namely, William R. Harper (1852-1906), a brilliant linguist. The university was so well conceived and funded that it had little difficulty in attracting as many outstanding scientists of the country to its faculty as it chose. To maintain status, other private universities were compelled to reexamine their attitude toward science and scientific research.

Stanford University was created at the same time (1891) in Palo Alto, California with the gift of an enormous parcel of land and enough endowment to get underway. Its first president, David Starr Jordan, was a biologist. He made certain, as funds became available, that the university put appropriate strength into its various scientific departments. In addition what is now the California Institute of Technology opened its doors at about the same time (1891). Its immediate goals were rather limited since it had the purpose of providing sound engineering education primarily to local students. The groundwork was very solid, however, and eventually permitted it to develop into a major scientific institution under the successive leaderships of Robert A. Millikan and Lee A. DuBridge.

Ten years later, John D. Rockefeller concluded that scientific medicine in United States was far behind its counterpart in Europe and that something should be done to stimulate basic medical research. As a result, he created what is now the Rockefeller University (1901) in New York City. At the beginning, it was primarily a postdoctoral research institution which gave great freedom to the scientists in charge of the various laboratories. One of its very important influences on the study and practice of medicine in United States was, at a critical time, to provide basic training in biomedical research to those who would become the leaders of the medical schools nationwide. Abraham Flexner, the brother of Simon Flexner, the first director of the institution described above, had great influence on medical education in the United States by advancing in a brilliant and convincing manner the doctrine that medical schools should be closely linked to research universities.

At the same time (1902), Andrew Carnegie created the Carnegie Institution in Washington, D.C. for general advanced research. It soon gained a worldwide reputation in the field of astronomy.

To summarize, as the United States moved into the 20th century, it had laid the foundations on both an institutional and educational basis to play a significant role on the international scientific stage. It had produced a few native born leaders and inherited some from Europe who were of world class stature and who earned appropriate recognition as such. Taken as a whole, however, the potentialities were much greater than the ongoing level of productivity. This situation was aptly described by Professor Henry A. Rowland (1848-1901) who held the chair of physics at Johns Hopkins University. When retiring as president of the American Association for the Advancement of Science in 1883 he made the observation with respect to the status of science in the United States that there were many weeds relative to the amount of grain in the fields of science. That there were grains of excellent quality, however, was demonstrated by the fact that in 1907 Albert Michelson (1852-1931) was awarded the first Nobel Prize to be received by an American citizen for his brilliant work in the fields of optics and astronomy.

To amplify matters further, one significant issue concerning the American university system should be mentioned here since it had a major effect on the development of the academic sciences, as well as the nonscientific disciplines, in later periods. The larger academic institutions eventually developed a system of departments in each major field, such as mathematics, physics, chemistry or biology, in which there were a number of faculty positions at each level of academic rank. This made it possible both to diversify the range of interests in a given field and to offer a broad range of lecture subjects and laboratory courses to students. It also made it possible to employ substantial numbers of graduate students (those who had already completed bachelor's work) as teaching and laboratory assistants. Such students in turn carried out studies and research for higher degrees at the masters and doctorate levels, thereby increasing the national pool of individuals having professional standing in their field. While, as Rowland indicated, the quality of most faculty and students in American universities may not have been uniformly very good by the highest international standards of the time in the 1880's, the system had the intrinsic merit that it permitted a rapid expansion of the cohort of scientists. Moreover, the average quality of those in the various fields improved over the decades as higher and higher standards were imposed by successive generations in the various professions. Fifty years later Rowland's comment would have been entirely inappropriate.

13. Industrial Laboratories

One of the particularly significant features of the early 1900's was the creation of a number of industrial research laboratories that worked in the area between the basic and applied sciences, making significant contributions to both. This practice first started in the electrical industry following a tradition that went back to Thomas Edison's laboratories in New Jersey. However, the policy was soon adopted by the chemical industry.

The first industrial laboratory which placed balanced emphasis on basic and applied research was that of the General Electric Company established in Schenectady, New York in 1900. Its mentor was a brilliant chemical engineer, Willis R. Whitney, who became its Director in 1904. One of its stars was the physical chemist Irving Langmuir who was awarded a Nobel Prize in 1932.

Taken as a whole the industrial laboratories proved to be extraordinarily productive. One of these, the Bell Telephone Laboratories, several of whose members won Nobel Prizes, is threatened at present as a result of the decision of a Washington, D.C. judge to break up the telephone company on the basis of what many regard as dubious socio-economic theory. It was formally established in 1925 to unify and extend research already under way.

14. Scientists in the Post-Civil War Period

The period between the Civil War and World War I brought to the surface a number of American scientists of world class stature. This indicates clearly that the underlying interest in science was present and that the overall opportunities to enter the major fields were improving slowly but surely. While many of the individuals spent a period abroad to gain experience the fact remains that their greatest work was carried out in United States. A few of the individuals are as follows.

J. Willard Gibbs (1839-1903)

Gibbs, mentioned briefly in Section 11 was an expert in the field of thermodynamics and did much to clarify the thermodynamic conditions under which discrete systems (phases) in contact with one another could be in thermodynamic equilibrium. He also reformulated along original lines the laws of statistical mechanics first developed by the Austrian physicist, Ludwig Boltzmann, in a manner that made them comprehensible and useful to a very wide audience of scientists. Gibbs was born and raised in New Haven, Connecticut, and spent most of his career at Yale University. He did, however, have a number of formative post-graduate years in Europe.

Samuel P. Langley (1834-1906)

He started his career as a physicist and astronomer with a focus on solar phenomena. After being made head of the Smithsonian Institution he became much interested in the science of aeronautics, carrying out experiments on the major forces on an airborne vehicle, such as lift and drag. This work was climaxed by the development of a pilotless steam powered model airplane which flew successfully and demonstrated that powered flight was possible. He also forecast the development of wingless transport vehicles (rockets).

Abraham A. Michelson (1852-1931)

He was an expert in the field of wave optics. Among other things he developed a very sophisticated interferometer which was used to determine whether or not light travelled with different speeds in directions parallel and transverse to the motion of the earth's orbit around the sun. The results were negative. He also made very accurate measurements of the speed of light and determined the length of the standard meter in France in terms of the wavelength of specific types of light. He had received his undergraduate education at the Naval Academy. As mentioned earlier, he was awarded the Nobel Prize in 1907.

Thomas H. Morgan (1866-1945)

He started his career as a research embryologist but became deeply involved in studying the role which the chromosomes and their component genes play in determining the characteristics of an individual. Using fruit flies as a medium of investigation, he and his colleagues demonstrated that the genes are of primary importance in determining the characteristics of a given fly. He was awarded a Nobel Prize for this work in 1933. It may be noted that some of the first world class experiments in basic biology carried out in the New World were in the field of genetics which was, so to speak, a relatively new area of research.

Ira Remsen (1846-1927)

Remsen studied advanced chemistry in Europe and subsequently made the Johns Hopkins University, where he received an appointment, one of the major national centers for training and research in chemistry. The advanced textbooks in chemistry which he wrote were translated into many languages and became international standards. He started his career with the intention of being a physician but switched to chemistry after gaining an M.D. degree and going abroad for several years to become familiar with European chemistry. He and a colleague discovered saccharin in the course of their research.

Henry A. Rowland (1848-1901)

Henry Rowland, a contemporary of Michelson, who was mentioned in Section 11, was in the second group of remarkable physicists to emerge in United States in the nineteenth century. After completing his basic studies at Rensselaer Polytechnic Institute, he went to Europe to gain additional research experience. Most notably, in the laboratory of H.L.F. von Helmholtz he carried out one of the most basic experiments of electromagnetic theory. He demonstrated directly with the use of a rotating wheel carrying electrical charges on its periphery that a moving electric field has a magnetic field associated with it and that the intensity of the magnetic field is proportional to the velocity.

On returning to the United States he accepted the professorship in physics at The Johns Hopkins University and became one of the international leaders in research in the field of wave optics. He is especially well-known for the development of ruling machines that could produce rows of narrowly and equally spaced lines on glass or metals. Such Rowland gratings can be used to break up a mixed spectrum of light into its component wave lengths.

He died prematurely of diabetes while at the peak of his career.

Charles D. Walcott (1850-1929)

He was an analytical geologist and paleontologist working with fossils in some of the oldest rocks available. Long before the period of radioactive dating of rocks, he devoted a large fraction of his scientific work in the search for fossils in the early stream of life and discovered the Burgess shales of Canada which possess fossils that date from early stages of the development of multi-celled organisms (The Cambrian explosion).[2] Walcott served both as head of the Smithsonian Institution and as president of the National Academy of Sciences. He had an exceedingly broad range of interests in science and the arts and a wide circle of friends both nationally and internationally. More will be said of him in the next section.

---

[2] In an otherwise excellent book, *Wonderful Life* (Norton, New York, 1989), Stephen Gould takes Walcott to task, not only for not making more in modern terms of this important discovery, but for accepting a set of values regarding the progressive nature of evolution which was not uncommon in Walcott's time. This is grossly unfair to Walcott. He was a pioneer in his field and he did appreciate the fact that the discovery he had made was very important, which is saying a great deal for him. Moreover, no one in the interval since Joseph Henry did as much as Walcott to advance the cause of good science in the United States.

15. The Influence of World War I

The outbreak of World War I in 1914 and its prolongation into 1915 caused great ferment within the United States for two good reasons. First, the United States tended to be cut off in part from much of European science and technology as well as manufactured products. Second, there was serious danger that it would become involved in the conflict. By 1915 the government began to call on the scientists and engineers for help as it began to establish policy. One of the results of this period of stimulus was the creation of the National Advisory Committee for Aeronautics (1915). After more than forty years of brilliant service it was transformed into the National Aeronautics and Space Administration (1958) which managed the Apollo mission to the moon.

In 1916 President Wilson asked the National Academy of Sciences to determine how the scientific and technical resources of the nation could best be mobilized in support of the country during the period of crisis. The Academy's response to this was the creation of the National Research Council with the help of industrial friends. The Council rapidly brought together the advisory services of scientists, engineers and physicians from government, industry, academic and private life on an extensive basis. The advisors were grouped into divisions spanning fields related to all areas of science, engineering and medicine of immediate interest to the government. The National Research Council soon penetrated the entire technological fabric of the nation and was of inestimable value both to the military services and to industry as well as to the executive offices. Its success was such that in 1918 President Wilson issued an executive order making the National Research Council a permanent part of the National Academy of Sciences. It has remained such. With the creation of the National Academy of Engineering (1964) and, soon thereafter, the Institute of Medicine, which are both closely linked to the Academy and serve as co-administrators of the National Research Council, the Academy has continued to be actively involved in most aspects of science, engineering and medicine in this country.

Two individuals who exercised a great deal of the responsible leadership in the creation of the National Research Council were George E. Hale, an astronomer who did much to promote American astronomy in the period between 1890 and 1930, and Charles D. Walcott, the paleontologist-geologist mentioned in Section 14. Hale served as chairman of the National Research Council during the critical years of World War I (1916-1918). The burden he carried during that period apparently helped to undermine his health for he suffered a serious breakdown soon after the War.

Walcott, in addition to being an excellent research investigator, as mentioned previously, was one of the most remarkable scientific administrators in national history. He was director of the Geological Survey (1894-1907), one of the initiators of the Carnegie Institution of Washington (1902), secretary of the Smithsonian (1907-1928), organizer and chairman (1915-1917) of the National Advisory Committee for Aeronautics and was president of the National Academy of Sciences from 1917 to 1923 which was very critical for its development since this included the time in which it acquired a home of its own. Walcott was very active in promoting matters vital for government-science relationships for a forty year period prior to his death. His method of achieving goals was quiet and subtle but enormously effective.

Many scientists went into uniform after the United States entered the conflict in 1917, hoping to increase their effectiveness in applied work and service. This included Dr. William Welch, the physician who was co-founder of the celebrated medical school at the Johns Hopkins University and who had been serving as president of the Academy.

The area of science and technology which received the greatest thrust from the War was chemistry. Prior to the War the United States imported a large fraction of its chemicals from Europe, particularly from Germany. As a result of the war, local industry was provided with the incentive to expand all phases of production. This was aided in part by the fact that industry gained access to material which previously was under the control of international patents. All phases of chemical research and production were stimulated and by the end of the War the industry was in a situation whereby it could enter into international competition. Academic education and research in the field shared in the benefits of this new environment.

16. Post-World War I

America prospered during the Golden Twenties. This prosperity led to the flowering of American science on many fronts. The Rockefeller Foundation, for example, created a new group of fellowships which made it possible for many new Ph.D. scientists to study abroad for one or two years and laid the foundation for an even more rapid evolution of science in American institutions. It would be difficult to overstate the role these fellowships played in accelerating the development of a research tradition. Robert A. Millikan and Arthur H. Compton were awarded Nobel Prizes in physics. American science began to approach the maturity of which Joseph Henry had dreamt half a century earlier.

With the surge of private enterprise, government involvement in scientific affairs remained fairly static whereas industry not only expanded its own research enterprises but, on a highly selective basis, provided research funds to university departments, particularly to chemistry departments.

Perhaps the most notable developments inside government agencies centered about the establishment of some military oriented research centers by the military services. For example, the Naval Research Laboratory was established in 1923 upon the recommendation of the great inventor, Thomas A. Edison. With the encouragement of the government the National Academy established the Highway Research Board (1920) which over the years generated extensive cooperation between federal, state and private organizations involved in the evolution of a truly national highway system.

In order to indicate that it now felt that its activities were beginning to be on a par with standards of world science, the American scientists stimulated the creation of the international scientific unions which assured regular international meetings in all major areas of science and encouraged the free flow of information not only back and forth across the Atlantic but also on a global basis. The unions have flourished ever since.

17. The Depression

The Depression years of the 1930's brought on a complete reanalysis of the role of the government in national affairs. It became recognized generally that, however great its virtues might be, private enterprise completely unaided by some degree of constructive national planning was probably not capable of exercising sustained leadership. Whether one liked it or not, our national government would find it necessary to play more than a passive regulatory role in relation to matters affecting the general public interest, granting that private enterprise had an indispensable and major role in the affairs of the country. There was a serious debate in Washington as to whether the greatest benefits to the nation would accrue from the systematic governmental support of science and technology or of sociology and economics. These were the days of the so-called Brain Trust and no one was quite certain what type of brain would be the most effective in solving the nation's problems. A group of engineers proposed a solution termed Technocracy which would in effect give the engineers jurisdiction over the national welfare. The social and economic scientists were also prepared to take over direction of the economy.

In this period President Roosevelt called upon the National Academy of Sciences-National Research Council to create an advisory board for a two year period (1933-1935), namely, the Science Advisory Board, which was invited to offer its services in coping with the national problems. Karl T. Compton, the president of the Massachusetts Institute of Technology, was made the chairman. The Board emerged in 1935 with a series of recommendations for government support of programs in science and engineering which look remarkably modern in retrospect. It also recommended that the science advisory service to the government which it was attempting to provide be a continuing one. Unfortunately, Congress was not sympathetic to such a program at that moment and the matter was shelved. In fact, some conservative scientists, such as Robert Millikan, strongly opposed any extensive governmental support of academic science and helped Congress to have the issue dropped at that time. Millikan probably was also guided by the fear that the "Eastern Academic Establishment" would dominate the distribution of funds—a not unreasonable assumption prior to World War II.

Fortunately, this short-sightedness did not extend to matters of public health for Congress did provide support for the creation of the National Cancer Institute in 1937 as a part of the Public Health Service. Since the Act establishing the National Cancer Institute permitted the granting of funds to private institutions, a mechanism had been devised for the vast expansion of activities in support of the public health which came after the War through the grants program of the National Institutes of

Health into which the Public Health Service evolved. In addition many universities received governmental support through the Public Works Administration. This made it possible to employ service staff, including students, for special work.

In spite of the drawbacks arising from economic restrictions, the depression years saw the American scientific community move ahead very rapidly in a number of fields which were to prove to be of great importance for the future. Growing activity in nuclear structure and radioactivity, accelerated in part by the discovery of the neutron in England in 1932, led to the invention by American physicists and chemists of new instrumentation for research in the field. This included the invention and development of such devices as the cyclotron and a great deal of auxiliary equipment. As was usual in American academic circles at this stage of evolution, any such interest evolved into a national enterprise with many geographically distributed "teams" essentially working both in tandem and in friendly competition so that numerous attacks were made on a given range of problems. Theoretical research in the field of nuclear structure advanced in close parallel with the experimental work. This practice made it possible for the American scientist to exploit the discovery of nuclear fission in Europe very rapidly once the news crossed the Atlantic in 1939.

In an entirely different area, research in the fields of organic and biochemistry blossomed. Industrial laboratories, working in close cooperation with academic ones, extended the knowledge of polymeric materials and created a number of new synthetic ones. Advances in biochemistry reached a stage which would make it possible in the immediate future (1944) to demonstrate that the structure bearing the genetic code was composed of DNA (deoxyribonucleic acid) rather than of proteins as had been long believed by many.

Not only were many native born and educated individuals attracted to scientific research in such a way as to swell the national cohort, but a number of brilliant scientists left National Socialist Germany either by choice or through dismissal and accepted posts in American universities. By the end of the 1930's the United States had become a leading scientific nation when measured by any standard. Seven American-born and three European-born scientists attached to American universities or industrial laboratories received Nobel Prizes in the fields of physics, chemistry and medicine in this decade.

18. World War II

The outbreak of World War II, much like the outbreak of World War I, caused an enormous and rapid transformation of the entire outlook of the country. Doubt about the importance of science and technology in relation to the public interest vanished almost overnight. Those individuals who had been active in the Science Advisory Board during the Depression and had recommended the creation of a more continuing advisory service came forward in positions of leadership and formed a National Defense Research Committee (1940). It in turn recommended the creation of the Office of Scientific Research and Development (1941) attached to the White House. This office set the pace for much of the national research and development effort in the course of the War, particularly that in which academic scientists were involved. Out of its dedicated and inspired work there grew not only a vast updating of the older disciplines of military research but also innovations that made an entirely new pattern of science-based warfare possible. For example, the Manhattan District of the Corps of Engineers, which exploited nuclear energy, grew out of the work of one of the divisions of the National Defense Research Committee. The wartime story has been told often enough that it need not be dwelt on further here.

From the standpoint of the postwar period, one of the most significant aspects of the contract programs developed under the government agencies was that it gave the working scientists of the country, particularly those in academic work, direct large-scale experience with major government support for the first time. American science was permanently changed by the experience.

19. Post-World War II

As the War neared its end, President Roosevelt called upon Vannevar Bush, the director of the Office of Scientific Research and Development, for guidance in determining the role that science should play in the postwar period once the wartime organizations had been dissolved. Out of this study grew the recommendation for what is now the National Science Foundation. Unfortunately the actual creation of that organization was delayed for six long years beyond the end of the War as a result of an extensive controversy that developed both in the White House and in the Congress. Key issues of conflict related to the way in which the director would be appointed and who would derive the benefit from any patents that might arise out of the work of the newly created foundation.

In the meantime, however, both the Navy and the National Institutes of Health took their own initiatives in following the patterns recommended by Bush's committee, gaining approval from both Congress and the White House. The Navy established the Office of Naval Research (1946) and the National Institutes of Health began an extensive system of research grants to nonfederal laboratories and health institutes. The Atomic Energy Commission was created in 1946 on the foundations of the Manhattan District and began its own contract program. The other branches of the Department of Defense soon followed the guidance of the Navy so that by the time the National Science Foundation was created in 1951 it was only one of several sources of funds for research support.

The European nations and Japan obviously experienced an enormous setback as a result of the ravages of war. The pursuit of science was greatly retarded, particularly in continental Europe. It became very attractive for foreign scientists to spend a period in, or even move permanently to, the United States. In the main, for a period of some fifteen or twenty years the United States became the leading scientific nation, exploring well-established fields with great energy and effect and opening many new ones. In the meantime, Europe, particularly the continent, has recovered its traditional commitment to all areas of scientific research and is using its resources very effectively following not only its own well-established traditions but taking advantage of some of the institutional procedures, such as large multi-user laboratories, first developed in United States. Perhaps a prototypical example of the latter is the great laboratory for nuclear and high energy research established by the European countries in the vicinity of Geneva, Switzerland. It is in successful competition with the best American laboratories working in the same field. The same can be said of the Laue-Langevin Laboratory in Grenoble, France, which is funded by France and Germany. It offers superb facilities for nuclear reactor-based research, including neutron sources, as well as facilities for research with magnetic fields of high intensity.

Since the end of World War II the United States has followed policies in the support of science that are somewhat different from those used by the European nations. Generally speaking, Europe has adopted an elitist attitude toward such support, funding as adequately as possible institutes or institutions at which there are outstandingly creative individuals working in significant fields of science. This means that those selected for support are treated well to the degree that the state of the ongoing economy permits.

While exceptional creativity is valued no less in the United States, there has also been a notable tendency, strongly advocated by the Congress which is representative of a democratic society, to spread the funds widely to competent individuals in many geographical areas. Unfortunately the number of good scientists has grown at a greater rate than the economy, including the fraction of funds the government is prepared to devote to science. To gain continuing support even the best scientists must spend a large fraction of their time documenting the case for it. This issue was not of great consequence until about the 1970's when available funds began to be stretched ever more thinly as the number of people engaged in scientific work grew at an essentially geometric rate. By the 1990's the issue had grown to what might be called crisis proportions—at least as far as the scientific community is concerned. It remains to be seen if it can be resolved effectively without taking drastic steps with respect to the reorganization of science in United States. Much of the future status of American science will depend upon the way in which the issue is handled. While private support for science continues to be important because of the great flexibility it can offer, there is no doubt that federal support of the right kind is indispensable. The way it is administered will be crucial for the future.

It is paradoxical that an academic system which proved to be so effective in expanding both the quantity and quality of those in scientific professions in a relatively short time in the national history should finally produce a major crisis in the fields of science emanating from the very factors which generated the successes.

Bibliography

A. Hunter Dupree, *Science in the Federal Government* (Cambridge Press, Cambridge, 1957).

Daniel J. Kevles, *The Physicists* (Knopf, New York, 1979).

E. R. Piore, *Science and Academic Life in Transition* (Transaction Press, New Brunswick, 1990).

Nathan Reingold, *Science American Style* (Rutgers University Press, New Brunswick, 1991).

5
Basic Science in the Asian Countries

One of the notable features of our time has been the extent to which nations on the eastern rim of Asia, particularly Japan, Korea, Taiwan and Singapore have adopted the methods of what, for the purpose of this chapter, I shall term western science-based technology for the manufacture of commercial goods which now flood the world in a highly competitive manner. Some of this production is carried out with the close cooperation of Western countries but increasingly a great deal is independent. It is probable that others in the Asian region may soon join them. Actually the Japanese began absorbing Western technology a century ago shortly after it was compelled to open its doors to world trade by the Western nations. However, much of the focus of such use in Japan prior to 1945 was for matters important to achieve military strength.

While the commercial successes of the leading Asian countries immediately after World War II were usually based on low-cost labor, most of the present success rests on their ability to produce goods of high quality with the use of very efficient manufacturing methods. Interestingly enough, those methods were thoroughly developed and documented by leading American industrial organizations, such as the Western Electric Company, at the time of World War I but were not used consistently in the United States in the decades immediately following World War II because of the absence of serious international competition—a situation which had changed abruptly by the 1970's.

1. The Issues

Granting that some of the Asian countries are becoming wealthy as a result of successful commercial enterprises based on Western science, one may rightly ask if those countries will also become major sources of basic science comparable to those found in Europe and the United States. This matter is important for two reasons, first the wider the range of sources of basic science, the greater the common pool of knowledge of the natural world for its own sake as well as for future technology. In other words, the larger the number of competent minds engaged in work at the frontiers of science the greater will be the expectation of new discoveries. Some of the very brilliant contributions to basic science that have been made by Asians working both at home and abroad leaves no doubt about the potentialities under favorable conditions. For example, Japanese scientists are well represented as honorary foreign members in many scientific academies outside Japan.

Second, should the Asian nations continue to "mine" the basic science of the Western countries without providing a comparable contribution to what might be termed the common pool, action may be taken in the West to cut back on its own contribution, particularly the segment having long range value, if supporting such research is regarded to be unprofitable. This issue has already become a matter of serious debate in governmental circles in the United States. In fact, a significant portion of the funds provided by the National Science Foundation in United States has been shifted to the support of engineering research that might well be funded instead by the mission-oriented agencies or, in some cases, by industry. This shift is made with the belief that the altered pattern will accelerate the output of technology of immediate value and thereby strengthen the domestic economy. Should this process grow, it will have a significantly negative effect on the creation of the pool of scientific knowledge that would ultimately be used, for example, in the next century.

2. Importance of Long Range Research

To appreciate the importance of long range basic research, it should be emphasized that although the transistor was invented at the Bell Telephone Laboratories in the period immediately after World War II, the invention rested on fundamental discoveries first made in the 19th century and which were pursued subsequently for many decades as topics of the most basic forms of scientific research. It was only after scientific understanding had achieved maturity that practical uses of these discoveries were found. Even then, knowledge gained from basic research in fields such as chemistry and metallurgy was needed before the original invention could evolve into the type of highly complex multipurpose integrated circuit system that has proven to be so revolutionary in the world of communications and information processing. Moreover, most of the basic research involved was part of a common international pool published in international scientific journals. While the interval between basic scientific discovery and its application may be much shorter today, and even more so in the future, the requirement that technology will need a broad scientific base will not change. One can hope the Asian nations will contribute to this base in reasonable proportion.

Nations such as Korea, Taiwan and Singapore are too close to their developmental stage to be expected to make major national issues of their contribution to the world pool of basic science at present. It is important, however, that they recognize the problem as they plan research budgets, particularly academic research budgets, both now and in the future.

As was mentioned in Chapter 3, India played a very important role in the early history of the evolution of science and clearly could do so again in the future. It, however, like several other countries in Asia, is dealing with economic and social problems which absorb a great deal of its vitality at the present time.

3. The Japanese Position

The situation is quite different for Japan which has generated a great deal of wealth as well as other important benefits by making use of the world pool of basic science and the associated technology over the past century. While it must be granted that Japan has been a major partner in the use of the world technological base for broad commercial purposes only since World War II, the issue at stake is whether or not it will eventually adopt policies favoring the support of basic science that are comparable to those which are being used in Europe or the United States.

Or will the Japanese system operate in such a way that those who are permitted to engage in basic research, for example, members of the academic community, come to be regarded primarily as adjuncts to a huge national industrial complex which is designed to generate monetary profits. At the present time it seems safe to say that the matter seems to hang in the balance.

That the overall situation may remain relatively unfavorable for the support of basic research is indicated by a number of factors related to the treatment of scientists in the Japanese academic community. For example, if a young Japanese scientist wishes to spend a period abroad in a postdoctoral post or on sabbatical leave, he or she can expect essentially no help from domestic sources except possibly from parents or relatives even though the funds required would be modest from the overall national viewpoint. In general, the young scientist must seek financial aid from the institution in the foreign country in which he or she desires to work. While most scientific institutions outside Japan that accept foreign students or young investigators are prepared to help visitors from indigent countries, in most cases, and from wealthy countries, in some cases, it is noteworthy that Japan enters into this form of scientific exchange as if it were an indigent rather than a wealthy country.

It is also notable that Japanese who have spent their scientific careers abroad and have received important awards for their research frequently express the view that it would have been far more difficult for them to achieve comparable success had they stayed in Japan because of what they regard as the constraints on those in the academic community.

At another level is the fact that in some of the most distinguished Japanese universities faculty of professorial rank are compelled to retire at about age sixty from the institution with which they have been associated and then receive very modest retirement benefits, ensuring a very frugal style of life. While they may then, and usually do, seek other positions, the change involved usually causes a great dislocation in their lives and that of their families since the post-retirement position, usually granted for a short time or on a year to year basis, may be at a place far from the individual's domicile. Housing cannot be shifted easily in Japan at the present time so a great deal of commuting may be involved. The stated reason for adopting such a system of early retirements is to make room in the research institutions for younger investigators, a policy often used in industrial organizations. The end result, however, is the denigration in status of the segment of the academic scientific profession engaged in fundamental research. The relatively hard-pressed life of the Japanese academic scientist does not provide an attractive model for the talented student.

It should be emphasized that the Western countries which are deeply involved in the conduct of basic scientific research face similar problems with respect to advancing young investigators. While there is no simple universal solution to the problem, it is usually handled in ways which do not denigrate the status of senior scientists who have made important contributions to the world pool of science. This should not be difficult to achieve in a relatively wealthy country.

It is true that at various times the Japanese universities, like others in the industrialized countries, have had student disruptions of sufficient magnitude to cause shutdowns for periods of time, thereby causing anguish to the society at large and alienating the citizenry to a degree. This problem can scarcely be solved to any significant degree through strict control of faculty tenure and auxiliary support, particularly in science and engineering.

4. Making a Decision

It is sometimes said that the authority for governance in Japan is very diffusely spread, that is, decisions are achieved by general consensus of the population in ways which are strongly influenced by what are termed long-standing cultural considerations. Accordingly, power is not primarily concentrated in the hands of government officials and industrialists, as often appears to be the case. Be that as it may, one can hope that some more equitable balance will be struck in the future whereby the current intense focus on monetary profits is compensated by the evolution of policies which assure a truly significant contribution to the world pool of basic science for the long as well as the short range.

6
The Future of Science

Let us attempt to look forward. As has been emphasized in preceding chapters, the two major influences upon society emerging from the evolution of science are conceptual enlightenment and technical innovation. Both are equally important in their ways. Science has shed a clear light of understanding on many aspects of the physical and biological world that were once deemed to be mysterious and potential objects for superstitious beliefs. On the technical side, it has provided radically new ways of improving classical technologies in fields such as agriculture, medicine, mechanical engineering and metallurgy. Still further and at least as important, scientific research has made it possible to generate entirely new areas of technology such as those linked to chemistry, electromagnetic devices, electronics, information processing and nuclear energy. All are essentially indispensable to modern life.

Beyond this, some aspects of science have their own intrinsic aesthetic appeal not unrelated to that of the arts, although usually appreciated principally by the dedicated professional workers.

For about five hundred years the attention and resources devoted to the pursuit and application of science have increased almost geometrically with time at great profit to society, not least to those in the industrialized countries. Along with this, as mentioned above, have been great improvements in health and longevity as well as almost miraculous advances in such matters as communications and travel.

Granting this, one may, nevertheless, ask about the future of science. Do we indeed face what has been termed an endless frontier and if so is it certain that scientific research and its application will be extended indefinitely?

It should be clear at the start that one cannot hope to predict the course of discoveries in any given field of science in detail, even under the best circumstances. One of the most important characteristics of scientific research is the continuous emergence of surprises—the natural world has its own inner structure and is not beyond turning our expectations on their heads. No generation of scientists can really comprehend in a detailed way where the continued broad pursuit of science will lead. In this sense our own generation is not different from those which have gone before. Discoveries to be made in the next century may eclipse all that has occurred up to the present and cause posterity to regard our status and outlook as somewhat primitive.

What is apparent in the present stage of evolution of science, along with increasing degrees of specialization, is the ever-broadening scope of the segments of the natural world, both physical and biological, which become amenable to investigation by the methods of science. Associated with this is the remarkable way in which disciplines increasingly intersect one another in spite of greater specialization by individual workers. Planetary studies become more deeply involved in geological and atmospheric science as well as basic chemistry and physics. Galactic astronomy leads into cosmology which in turn intersects with high energy particle physics. Geology, oceanology and atmospheric science become linked to studies of holistic earth dynamics instead of remaining descriptive studies of relatively static systems. Reductionist aspects of biology call upon highly sophisticated chemistry and require the use of complex physical instrumentation. Medical science finds that some of its most revolutionary advances borrow discoveries from the work of the cellular and molecular biologists. Some branches of mathematics can advance most effectively only with the use of super-computers. Computer design, in turn, relies increasingly on advances in chemistry, physics, materials science and mathematical logic. The field of neurophysiology which is still in its infancy promises to develop cross links with complex aspects of the theory of information processing as well as cellular and molecular biology.

The vast extension of chemical knowledge that emerges from the field of biochemistry assures us that aspects of science associated with the preparation and study of synthetic materials for their own sake or for ultimate use is still in its infancy. For all practical purposes there is in this area, as elsewhere, an essentially limitless frontier.

1. Factors Favorable for the Continued Advance of Science

Innate Curiosity

The most important factor assuring the continued advance of basic science lies in the combination of earnest curiosity regarding nature and the desire for self-expression that resides in many talented and imaginative young people—deeply ingrained human traits. These traits have been instrumental in the evolution of science from its beginning. Indeed, such curiosity can continue unabated for a lifetime in the well-initiated, not least in the professional scientist, in spite of varying levels of creativity.

Alongside this we now possess, as a result of nearly five centuries of experience, knowledge of the combination of experiment, logical analysis, speculative theory and institutional structure needed to form a solid platform for the advance of science. None of these guarantee the appearance of that flash of inspired insight from a great mind that is occasionally necessary to introduce a major new evolutionary concept in some field. In this respect we will apparently always depend upon the arrival of the appropriate level of genius at the active scene during special periods in the development of a field. Fortunately, such pregnant moments seem to attract the appropriately gifted sooner rather than later. One can only hope that this will continue to be the case indefinitely.

Practical Need, National Pride

Also on the positive side, it seems clear at present that under normal circumstances the advanced industrial societies will have a continuous need for the further infusion of new scientific knowledge for several good reasons. Some of the need will arise from a basic interest in the revelations of science, some from its educational value, and some from issues such as the improvement of public health, industrial competitiveness, defense and what might be called replacement technology—such as finding substitutes for materials in dwindling supply.

Then too, there is national pride, which has been a significant motivating factor in the past and which will probably be significant as long as we have a diversity of ethnic and cultural groups on an international scale. Doubtless national pride as well as an interest in science played a role in the formation of a number of national academies of science at the same time as the Royal Society of London in 1660.

Global Issues

Finally, it is likely that there will be global problems that require the encouragement of reasonably coordinated basic as well as applied research at many centers on a world-wide basis. Issues such as concern about the global environment or problems related to health such as cancer and acquired immune deficiency syndrome (AIDS), not to mention as yet unforeseen but inevitable pandemics, will require enlisting scientists from many institutions who are prepared to work at the most basic levels of current understanding. It is, in fact, remarkable that the worldwide epidemic of AIDS occurs just when our ability to achieve understanding of the disease is possible and when detailed scientific investigations can be carried out internationally in a concerted way. This is undoubtedly not the last time that the international scientific community will be called upon in a similar manner.

Then too, there will be less life-threatening scientific adventures which can benefit from international cooperation. The coordinated research programs in the antarctic provide one present-day example. The development of very high resolution astronomical observatories on the moon, including extended arrays which might observe planets on neighboring stars, could provide such an international adventure in the near future.

2. Extension of Scientific Research

If we grant that the continued advance of basic science is exceedingly important, indeed indispensable for the future of civilization, we may well ask whether or not such pursuit can be expected to occur more or less automatically on a worldwide basis, much as it has throughout Europe and North America. It should be emphasized that we refer here to the development of the basic rather than the applied sciences which are far easier to transfer from one culture to another.

The extension of research in basic science in any given society will depend, among other things, on the level of intellectual freedom and the availability of institutions which can provide the base and create the structure for a discussion and review of scientific issues. It will also require a minimal degree of wealth, the pursuit of some fields requiring more, others less.

It is notable that within the centers in which basic research is thriving individuals from many other cultures and with a great variety of ethnic backgrounds have been able to make substantial, even brilliant, contributions. The important requirements are that the individual have a deep personal interest in science, a willingness to make a serious commitment, innate talent and the ability to pursue close communication in one way or another with those in the frontier sections of the field in which he or she decides to work. Most communities in which the scientific culture is deeply imbedded can and have served as hosts for individuals from many other societies when the requisite desire, interest and talents are available.

Although individual talent is broadly distributed around the globe it does not follow that all cultures provide equally good environments for the production of truly creative basic research even when the wealth and scholarly institutions associated with such cultures are in other ways remarkable. The path to a more nearly worldwide extension of sources of basic science may be slow and possibly difficult to achieve. Let us consider some examples.

We have already discussed such issues relative to the wealthier countries in the rim of East Asia in Chapter 5.

To date Mainland China has not been able to become adequately integrated into the world scientific community although it has a long history of outstanding cultural and technical development. This, as we have seen in Chapter 3, actually transcends the circumstance that it currently has a government that is an oppressive dictatorship which views its intellectual community with suspicion. There have been long periods in the history of Chinese culture in which the political situation was otherwise. It will be interesting to observe the developments which take place in Taiwan in the period ahead since it is encouraging the advance of science with the means available to it.

Japan provides another interesting case. Although its current economic success is based to a major extent on the acquisition and use of borrowed science-based technology, it cannot be said to be making comparable contributions to basic science commensurate with its wealth and technical skills. As mentioned in Chapter 5 one suspects that the society may have chosen to adopt policies which do not encourage the pursuit of basic science to the extent found in most European countries, placing instead maximum emphasis on matters of immediate practical consequence. Such policies could, of course, be modified but it would, at a minimum, require a broad reexamination of the framework within which the scholarly community, or at least the scientific community, is supported.

To repeat a point made in Chapter 5, India was a major contributor to the advancement of science in its early history, well before the rise of Islam, through creative work in mathematics (analysis) and astronomy. One suspects that it could readily become a major contributor again if its current economic and social problems were resolved. Indian scientists like their Chinese counterparts have performed brilliantly in modern times when the cultural environment has been appropriate.

Finally, let us consider the Moslem world which once extended from the Atlantic Ocean and well into Asia. As we have seen in Chapter 3, it played an enormously important role during its peak centuries, say between 700 and 1400 A.D., by consolidating the scientific knowledge of the known world, ranging from India to the Mediterranean, and placing its own remarkable imprint upon the fusion. Moreover, Moslem scholars through cultural transfer sparked the Scientific Revolution which emerged in Europe in the late Middle Ages and Renaissance.

Unfortunately, Islam has suffered setbacks in the intervening period which have caused its intellectuals to retreat to a considerable degree from the scientific stage. It remains to be seen if this trend can be reversed.

3. Factors That Could Lead to the Decline of Science

There are a number of factors that could lead to the decline of new activity in science. Fortunately most, although not all, are probably limited in scope and effect when viewed from a world-wide perspective.

Decline of Interest

The most obvious is a possible decline of interest among those with the greatest talents. This seems unlikely if we view the challenges offered across the frontier of science. One can imagine that some fields such as classical nuclear physics or even high energy particle physics eventually could become less attractive to the most brilliant and creative individuals as the opportunities for new developments or applications appear to be exhausted. Actually a decline in interest in high energy physics seems to be relatively far off, when viewed at present, in as much as the basic theory surrounding high energy particle physics remains far from complete and there is such a strong link between that field and cosmology which remains a very popular subject.

Similarly, the challenges offered by the life sciences seem to be essentially unlimited at the present time, particularly if we consider our still primitive understanding of cellular mechanisms, including the very important matter of cellular specialization, and the unresolved intricacies of the nervous system. Then too we have only the most rudimentary understanding of the interrelation of the holistic and the molecular properties in even relatively simple biological systems—a topic which was once deemed central to the subject, but is now almost ignored by "main line" biologists. We will return to this matter in Chapter 8.

Public Interest

It is evident that communication between the creative scientists and the general public is becoming more and more difficult as science advances in both sophistication and specialization. While this may well weaken the everyday involvement of the average person in the more profound aspects of science, there undoubtedly will be continuous and widespread attention to the practical fruits of science to the extent that they have an effect on everyday life. In spite of this it is very important that those individuals who are active in encouraging the continued public support of the most basic frontier aspects of science develop and use whatever means and talents they have to retain public interest. It will inevitably have an influence on the magnitude of such support.

Student Involvement

It is well-known that the dedication of second and third generation students to fields of science has dwindled in the United States—somewhat reminiscent of the way in which involvement in professional engineering dwindled in the United Kingdom in an earlier generation. This is a sociological matter of complex origin stemming, among other things, from a shifting sense of social values—that is, in what is important in the life of the individual. Doubtless one of the factors causing the decline is the decrease in the quality of the teaching profession at the primary and secondary levels and its inability to strike the cords of inspiration in the students. Fortunately for the United States, devotion to science has not diminished among the children of recent immigrants, particularly those from Asia, who seem to have their own sources of inspiration. Moreover, the commitment to science in continental Europe and the industrialized Asian countries appears to be maintained at least on a continuing level granting that the interest tends to be on short range aspects at present in the second case.

Cost of Equipment

It is hardly necessary to mention that the cost of equipment and related support for research in some of the most interesting areas of science has become increasingly high as the problems have become more and more intricate—a topic to be discussed in Chapter 12. As commented earlier, there are even branches of present day mathematics which rely upon the use of the most advanced and expensive computers. Biological research has long since passed far beyond the string and sealing wax stage. Several million dollars is needed to equip even a modest biochemical laboratory for frontier research.

It is not inconceivable that the pursuit of some areas of experimental physical science, such as high energy physics, will eventually require quite different approaches from those used at present. It is reasonable to suppose that the super-conducting super collider (SSC) now being planned for the State of Texas will represent the last in the evolving generation of particle accelerators which started about sixty years ago with the development of accelerators such as the cyclotron.

One should not, however, underestimate the ingenuity of the imaginative scientist in finding less expensive, although perhaps more time-consuming, ways of achieving challenging goals. In some cases such ingenuity even leads to great advances. While astronomers still use large mirrors for their research, ingeniously designed phased arrays of receivers are providing valuable astronomical information that would other-wise involve exorbitant costs or be completely beyond reach. Similarly, it is quite possible that space platforms and space vehicles developed for other reasons will in the future provide the means for research in areas of basic science which would otherwise become too expensive to pursue. Cosmic ray research carried out beyond the atmosphere may become central to high energy physics in the future.

It seems unlikely that limitations placed upon the budgets for science will impede its progress in any absolute way as long as inspired and imaginative scientists remain active, even though the approaches to some areas of investigation may require different methods of attack. What is important for the continuation of progress is that some more or less well-defined fraction of the wealth of the industrialized countries continue to go to the support of relatively free research and that the expenditures be determined primarily by those familiar with basic sciences as described in Chapter 12.

It seems reasonable to assume that the fraction of national wealth expected to be devoted to the most basic, relatively unrestricted form of research is somewhere near its maximum at present in most industrialized countries in the Atlantic area. The fraction can, however, be expected to vary somewhat from time to time and from country to country depending upon circumstances. In any case there will inevitably be limitations on the opportunities available to individuals interested in the most basic forms of research. Such limitations will be much less severe in areas of applied research and development.

In summary, it would appear that only a major violent sociological or physical upheaval of an essentially global nature could change the general advance of science, granting that the areas to which major attention is directed at any given time may vary as a result of fashion, economics or an urgent public request based on some issue of real or perceived importance for the well-being of society, for example, public concern with respect to acquired immune deficiency syndrome (AIDS) or access to new energy sources.

4. Effects of Anti-Science Movements

If we look back upon the history of science, we recognize instances in which the course of science was arrested or even suppressed as a result of social pressure associated with a powerful orthodoxy. Aristarchus' proposal of a solar-centered planetary system was rejected as unorthodox or even heretical at a critical moment in the evolution of Greek science. Indeed, were it not for the fact that large portions of Greek science found an enthusiastic reception in the Islamic world and became imbedded in its culture as an active ingredient, the foundations generated by the Greeks might well have been buried in the ruins of antiquity. Turmoil in the early Christian world, as it sought consolidation, might have proved fatal to the resumption of the development of science.

Then too, in more recent times, the Inquisition attempted to arrest the advance of science as part of its program to suppress the Reformation and other "heresies." Fear of its authority and power were sufficient to delay the publication of Copernicus' great treatise, just as it was sufficient to cause Galileo to deny publicly the truth of his firmly held beliefs. In still more recent times, we have had cases in the Soviet Union, hopefully now on a different track, in which well-established areas of good science were suppressed. Among other incidents was the arrest and exile from Moscow to Georgia of George Gamow in the late 1920's for lecturing openly in Moscow on the newest developments of quantum mechanics. The head of the institute at which he lectured was sent to Siberia. There is also the authority bestowed upon Lysenko in the 1930's which enabled him to suppress work in the field of Mendelian genetics and which led to the premature death of Nikolai Vavilov.

Alongside this was the spurious distinction made between so-called Jewish and Aryan science, built into a brutal and fiercely destructive doctrine, by the National Socialist regime in Germany. It led to a great decline in science there for a long period of time.

The Scopes Trial in Tennessee and the impediment of anthropological research in South Africa during the period of Apartheid are peripheral and fortunately limited examples of threats to the natural evolution of science. Whatever else, the Creationist movement in United States must be regarded as a continuing, if minor, threat to the pursuit of science. Its principal influence is to cause distortions in elementary education in the sciences.

Had the Soviet Union ever achieved the type of world hegemony that was pursued by Gorbachev's predecessors, one might wonder if its leadership would have continued to pursue scientific research with enthusiasm. The fact that the Soviet academy of sciences could be placed in a position whereby its membership felt it necessary to endorse the demotion of Sakharov says a great deal about the previous leadership's attitude toward science and the scientist. The uncompromising attitude of the present government in mainland China toward its intellectuals speaks volumes about the continued fragility of science in that country. Those who saw that the university and other basic research institutes in mainland China had been reduced to shambles as a result of the Cultural Revolution can have no illusions concerning the stability of science policy in a dictatorship.

All of this reaffirms what has been said many times previously: The best guarantor of the advance of science is a social and political structure which places a high premium on openness, the encouragement of free enterprise and minimizes centralized control of society to the extent feasible, granting that the government must retain sufficient authority to promote the health, freedom and prosperity of its citizens.

5. Virtue and Evil in Science and Technology

As has been mentioned earlier, there have been various times in human affairs in which individuals or groups have risen up to propose a halt to technical progress—the so-called Luddite response. In recent times the objection has been extended to almost any area of science that might lead to new technology.

In some cases the proposals are well meaning in the sense that they are intended to come to the aid of a special group being disadvantaged by new technical trends. The political actions taken against the use of milk-producing hormones in some regions in United States and in other countries which have many small dairy farms provide an example of this.

In other cases, many of the objectors seem to have much broader and diffuse complaints, being perhaps personally bewildered by the pace of change and desiring a complete restructuring of society—as if science and technology were directed in some way by forces of evil intent.

Fields of investigation such as those focusing on peaceful and safe uses of nuclear energy are frequently attacked broadside these days as is described in Chapter 10. In the decade after World War II and in the wake of Robert Oppenheimer's declaration that the physicists "had known sin," some biologists tended to frown upon those involved in the physical sciences as if their own activities were so obviously humanitarian in spirit as to be utterly beyond objection. The opponents, however, have since disillusioned them by attacking aspects of recombinant DNA research, a natural outgrowth of brilliant advances in cell biology and biochemistry.

Science must follow its own channels, adding to our store of available knowledge and understanding. The use to which society places this knowledge represents activity at an entirely different, more socio-political, level in which circumstances, as well as ethical and moral considerations, enter the picture.

The disease cancer is universally regarded as an evil. No reasonable individual would question research which seeks to discover the factors either inherent in organisms or in the environment which can lead to cancer even though under certain circumstances that knowledge could be misused to induce cancer that would not otherwise occur.

Ever since mankind developed tools, learned to control fire, and began the domestication of animals and plants, the knowledge our species developed could be put to a variety of uses which might, in appropriate situations, be regarded as either good or evil. Knowledge of the inherent capabilities of our discoveries or inventions is morally neutral. Ethical issues emerge when actual use is contemplated. Here the matter may become highly controversial. One is reminded of President Truman's response to the issue of sin after a discussion with Robert Oppenheimer: "He thinks he ordered the use of the bomb."

Let us raise an admittedly hypothetical and complex issue which the scientific community could eventually face if our understanding of human psychology advances sufficiently—an open issue in itself.

The past century, and indeed many earlier centuries, have provided instances in which charismatic leaders, using grossly empirical techniques in an intuitive way, have succeeded in inducing otherwise decent individuals to engage in activities which turned out to be highly destructive to others, or to themselves, or both. In most such cases the initial intent of such movements may have seemed rationally and even ethically justifiable. However, they clearly can take on a life of their own in a way reminiscent of the development of a cancer and become highly destructive.

The question arises: would it or would it not be morally appropriate to attempt to understand this aspect of human behavior at the deepest possible level of scientific investigation? Assuming success, the knowledge clearly would constitute a double edged sword. In the best conditions it could, in principle, be used to forestall dangerous political and social movements. In the worst, it could strengthen the powers of an unscrupulous dictator.

The basic issue here, as in many other cases, has much less to do with the morality of scientific knowledge than it does with the principles upon which a given society is based and the safeguards it adopts to forestall dangerous excursions linked to the misuse of power.

6. Trends in the United States

A segment of the anti-science movement in United States deserves special mention since it shows that an abundance of freedom in a society may not alone guarantee the unhindered advancement of science. Quite apart from the Creationist movement mentioned earlier, we see, particularly in the United States, the rise of what could readily become an anti-science movement within substantial portions of the intellectual community, not least the university linked community. Aspects of the opposition to the continued free advance of science and technology that are related to concerns about the environment are described in Chapter 10. At its worst this movement is driven by groups of individuals who would seem to prefer a much more restrictive society in which there are far more centralized controls and in which further technological advance is either brought entirely to a halt or is subject to great limitations. Such groups tend to be well-funded from both public and private sources and to receive a proportionately large amount of favorable attention from the media. Scientific research, as one of the principal sources of new technology, should, in the view of such groups, be subject to close and not as yet well-defined controls. A significant portion of the attacks on research in the field of recombinant DNA mentioned earlier are linked to this anti-science movement. In its own way, it joins forces with the Luddites.

What is most disturbingly contradictory about this trend is that imbedded in it is the failure to recognize that it is the wealth and the advances in technology gained through the free workings of science in the past few centuries that have enabled the intellectual institutions in the United States (and incidentally elsewhere) to thrive and support a relatively large fraction of the population in intellectual pursuits. It is paradoxical that the activities of such anti-science groups, if successful, would ultimately undermine the very economic and social structure which supports them. One can only imagine that the activists feel that they would be immune personally from the burdens experienced by humanity in the pre-scientific age in which life for most individuals was poor, short and nasty.

7. Barriers Originating from Within Science

In conclusion, let us, for the sake of discussion, assume that an interest in the pursuit of scientific research continues and that freedom and resources are made available for it. What then can be said about obstacles that might arise out of the natural course of the evolution of science itself. We shall here leave aside issues related to the increasing cost of experimental equipment in some areas of science.

There are two possible obstacles that deserve mention. First it is possible that in the course of advancing given areas one may encounter levels of complexity that are beyond our capacity to deal with in any direct way. For example we may face issues which are beyond treatment, even statistically, on the basis of knowledge gained from reductionist analysis—the tool which has been so powerfully useful since the dawn of science. We may never be able to use all the information being accumulated by molecular biologists to understand the workings of a living cell in a detailed or even qualitative manner, as indicated in Chapter 8. Accurate long-range predictions of the weather or of earthquakes may turn out to be hopelessly complex. If, or perhaps when, this occurs, we may be compelled to adopt new attitudes toward the process of gaining understanding of nature, with a fresh appreciation of our limitations, whether for human or more cosmic reasons.

Second, we undoubtedly will also face imponderable issues which are entirely beyond our comprehension, let alone our ability, to treat qualitatively. For example, can we ever even begin to understand, through the methods of science, the wherefore of our universe with its intricate structure even though we may trace its origin back to an initial cosmic explosion? Will we ever comprehend the qualities of the human mind both in depth and in totality beyond knowing the details of its molecular structure and the simpler forms of molecular interaction?

7
The Next Million Years—
A Half Century Later

Just after World War II, there appeared an ambitious book with the title, *The Next Million Years*[1] Its author, Charles Galton Darwin, was a distinguished physicist and the grandson of the great Charles Darwin, author of *Origin of the Species*. The younger Darwin, then in his early sixties, had not only had an outstanding research career but had travelled the world, served as liaison representative of the United Kingdom with the White House in World War II and had also been director of the National Physical Laboratory in England, an institution central to much of Britain's research during that war. As was his privilege, Darwin, a Malthusian at heart, came to the pessimistic conclusion that the long-range future of humanity is not very bright. To many, this conclusion seemed overly gloomy since we had just emerged victoriously from a challenging war and the world seemed like a much brighter place.

First we shall review here the core of the arguments upon which Darwin drew his conclusions and second present, from the present author's own perspective, the situation as it can be judged a half century later, granting that Darwin's pessimism still has much to support it.

1. Choice of Time Period

Darwin chose to look forward for a million years for two principal reasons. First he felt that it would require a period of at least several hundred thousand years for the basic genetic makeup of our species to change significantly. In this connection he doubted if there could ever be general agreement about how it should be changed other than perhaps for genetic diseases if mankind indeed did succeed in finding ways to alter our genes. Second, he clearly wanted sufficient elbow room to counter critics who might

---

[1] Charles G. Darwin, *The Next Million Years* (Rupert Hart-Davis, London, 1952).

raise objections to his overall conclusions with the use of what he would regard as temporary, short-range fixes. For example, he wanted to cover the possibility that attempts would be made to avoid his Malthusian prediction by providing what he would regard as temporary solutions. He believed that an inevitable geometric growth of population would ultimately outrun any new source of food that might be generated.

One might be inclined to focus on a period of time much shorter than a million years since many of the issues which most concerned Darwin have already been subject to changes in recent decades in ways that open at least a ray of hope for the long range. Moreover, in view of the present pace of societal evolution, one might expect the fate of humanity to be sealed in much less than a million years. Darwin probably felt the same at heart.

2. Statistical View

Darwin was much impressed with the works of Arnold Toynbee which were appearing when his book was written. To a degree Toynbee's views fitted in with his opinion that one could hope to predict the consequences of human behavior, when examined in the large, by what might be termed gross statistical methods. It will be recalled that Toynbee, who possessed a remarkable knowledge of world history, had extended earlier theories of Oswald Spengler. The latter had proposed that the development of any given human society would follow a more or less universal trend, eventually leading to stagnation and collapse. Toynbee believed that a society could avoid a given trap leading to stagnation if, at such a time of crisis, it could muster the vitality and imagination to cast itself into an appropriate new mold and overcome the crisis. The transition to the new state might well involve a substantial period of ordeal and a major shift in social or political values. Roman civilization, for example, twice avoided the threat of stagnation by extending citizenship first to all the tribes in Italy and then to members of the entire empire. It also overcame another serious crisis by accepting Christianity as the official religion.

Darwin had been a significant innovator in the field of statistical mechanics and had a well-developed sense of the power of statistical methods if one is willing to focus on large-scale rather than microscopic details. He was prepared to ignore such matters as natural or geographic boundaries or ethnic differences and concentrate on the most universal human attributes. He was also prepared to ignore matters such as global climate change, assuming that humanity would adjust through shifts in the location of major populations if need be.

3. Principal Human Characteristics

Darwin emphasized two human characteristics that he regarded as centrally important for his conclusions. He also called attention to a third which I believe should also be given major attention in a critique of his book.

Wild Animal

First he observed that the human species must be classified among the wild rather than the domesticated animals in spite of the level of sophistication its civilizations may have achieved. Domestic species have been bred for specific features under the close supervision of breeders or masters who determine their fate. The human species has never been subject to any such direct supervision under an external master or breeder. It is responsible only to itself. Our genetic characteristics and behavior are, to a large extent, determined by a combination of chance, environment and social development. When a segment of humanity does fall under the control of human mastership, the individuals who serve as masters are selected in one way or another through human intervention. The mastership is in turn not controlled from above. Any such selection represents a transfer of the freedoms of a component of the wild species to one or a group of its members who are in turn themselves free of control.

Population Growth

While some membership groups of the human family may learn to maintain the size of their population well within the limits of their food supply so that there is ample food and related amenities for essentially all, other groups of people tend to multiply at a rate which presses their population to the limit of the supply of food, thereby achieving a state in which the quality of life and civilization become depressed. Darwin felt that the improvement in modern medicine resulting from the introduction of newly discovered agents such as the antibiotics, which can decrease mortality among infants and older citizens, would tend to accelerate the growth of populations and thereby exacerbate the effect. He also believed that it is inevitably the fate of the human species to be dominated by the groups with large and continually growing populations. This trend should in the long run reduce the living standard of humanity below that required to provide a continuously advancing civilization. In brief, he believed that it is the fate of mankind to end in a universal state of impoverished stagnation. Inherent in this was the belief that it will prove impossible for geographically separated segments of the population who do live within their food supply to prevent being overwhelmed by those who do not.

Creeds

Incidental to his analysis of what he regarded as invariant features of human behavior, Darwin noted that population groups can, essentially on their own volition and without being forced either by dire necessity or by the will of the leadership, develop what he termed "creeds." These, so to speak, produce behavioral patterns which are widely and willingly accepted within a given group and which may have profound consequences for the group in the long-range. Such creeds include religious beliefs but need not be linked to them or to governmental edicts. One excellent example is provided by the deeply held view among the Chinese that each individual has a personal responsibility to provide a link between the past and future by having heirs—in the Chinese case preferably male heirs. While this creed may have found its origin in a state of affairs in which the aged were almost certain to face destitution if they did not have children who could support them in their later years, the creed has established a life of its own. As long as the practice persists the Chinese gene pool will be preserved, granting that it faces the hazard of producing overpopulation, which we appear to witness in Mainland China. Some individuals have felt that the absence of an analogous creed among the leading Romans was a significant factor in their decline.

It is also interesting to note incidentally that at the folk level the Chinese do not believe that one's religious views should be a source of contention but are, instead, to be regarded as a private, personal matter. The country has not known internal religious wars in historical times although there may be external pressures that are partly religious in origin, as from the Moslems in the west.

4. The Four Stages of Civilization

Darwin recognized that thus far the civilization of mankind has passed through four successive stages of evolution, namely, those based on the use of fire, the development of agriculture, the development of urban life and the use of basic science for technological advancement. Two important comments may be made with respect to these four stages. The first three developments undoubtedly came about as a result of pressures at the survival level. As a primitive wild animal, man must have had the same innate fear of fire that is common to most animals. Presumably he found, however, that under dire circumstances it could offer him protection, comfort and other benefits in spite of his initial fear. As a result, he decided to use this mysterious agent of nature as an adjunct to life, eventually expanding enormously on its use without really understanding it. Similarly the able-bodied male hunter must have been in desperate straits to help feed his tribal group before he would be willing to join the women and others confined to the camping ground in the labor of nurturing domestic plants and animals. Doubtless village compounds were developed initially to serve as protective centers once agriculture forced the tribe out of the cave and into relatively open areas.

5. Science-Based Technology

The origin of science-based technology deserves special discussion. As was emphasized in Chapter 2, mankind has depended upon technical innovations for some of its essential needs from the dawn of civilization, and has always cultivated what might be termed engineering talents, using knowledge gained through experience and everyday observation along with the inherent ability to invent. Science itself entered through the back door so to speak as a result of the curiosity and systematic experimentation carried out by a highly specialized group of individuals who decided to look ever more deeply into the relationships to be found in the natural world about us. Some of these individuals had, incidentally, the aptitudes of superb engineers but the basic motivation of the group arose out of natural curiosity, not the desire to solve practical problems.

The discoveries emerging from such basic research were initially of only slight use to the engineer in his everyday work. While the development of optical science in support of the production of spectacles may provide a landmark case, one of the first truly major uses of accumulated scientific knowledge other than for intellectual enlightenment may, as we have seen, have occurred in the university established in southern Portugal by Henry the Navigator as he sought to promote the exploration of the planet and establish new trade routes to Asia. One must wait almost until the 19th century before the discoveries of basic science began to create entirely new profitable industries on a large scale.

The fact that the well-springs of technology derived from basic science are, so to speak, more deeply subterranean than are the foundations of more obvious forms of technology means that the conditions needed for the nurturing of these well-springs is not clearly understood by most people—including most engineers, who serve society much more directly. It remains to be seen how well this gap in understanding can be bridged in the long term.

6. The Fifth Stage of Civilization

Darwin reached the conclusion that the four stages of civilization would inevitably be followed by a fifth that would climax the fate of our species. It would be a period of great deprivation. First and foremost, we would have exhausted the supplies of energy derivable from fossil fuels. Along with this we would have depleted the earth of the richer sources of essential materials for industry and agriculture. In the meantime the human population would have grown to such a high level that the contraction of resources would necessarily lead to dire poverty for all humanity. Freedom of action for the individual or group would be essentially lost. The picture which emerges is, to use present day language, of a global third world environment in which the prospects for advancement to a higher level of well-being are negligible because of constraints. This is the aspect of Darwin's book which led to the statement that his outlook for humanity was indeed very gloomy.

7. Bullies and Birthrates

Darwin had two other principal concerns. First he noted that humanity normally develops a certain fraction of individuals who are inclined to be bullies, that is, who wish to dominate others by force, extending to brutality if necessary. Normally bullies are kept in line by the rules of societies in the most politically advanced countries. Dictators emerge from the ranks of bullies and they usually have little difficulty in gaining a following of individuals of like temperament. They can be a source of instability to any society in times of stress. What they may lack in numbers, they compensate for by belligerence and brutality. It would appear that a well-functioning democracy is the best antidote to the rise of bullies to dictatorial power, but some form of international cooperation to forestall their activities may provide a useful adjunct.

Second, he was concerned with the fact that in some of the most industrially advanced countries, those most capable of providing leadership, innovation and productivity normally have fewer offspring than the less capable as a result of the desire of the former to maintain a fair standard of living and good opportunities for their families. He felt this trend would, in the long run, lower the median level of intelligence and place mankind in a position where it is less and less able to deal with the problems it faces. Darwin, incidentally, placed intelligence at the top of the list of attributes humanity had gained through many millennia of evolution.

8. Prelude to Critique

Before attempting anything that might be called a detailed critique of Darwin's thesis, one must admit that in a broad sense he has a very strong case—the odds weigh heavily on his side. A review of the events in the 20th century with its two enormously wasteful world wars, relentless growth of population in the less developed countries and endless violence does suggest that our species is headed toward some form of doom of its own making. And yet we must also admit that the situation is not utterly hopeless. As individuals, the members of our species have a desire to survive. That drive has been effective in the past even though success has often been touch and go and has occasionally required that individual societies make strong breaks with attitudes and patterns of behavior accepted in the past.

To bring matters up to date, we may undertake a critique of some of Darwin's basic assumptions as they can be viewed fifty years later and then suggest a fairly stringent list of basic conditions required for our long-term survival, granting that the obstacles are formidable.

9. Nuclear Energy

In spite of the fact that nuclear energy was available at the time Darwin's book appeared circa 1950, he dismisses the likelihood that it would be available for general use in civilian life. This conclusion is undoubtedly based on the assumption, commonly held for a short time during and immediately after World War II, that uranium is a very rare element and that the little available would end up in military hands. In fact, the author recalls a lecture given by Enrico Fermi in the winter of 1944-1945 to the effect that uranium was so limited that the military would have to decide whether to use it to make bombs or for ship propulsion. Explorations soon proved, however, that uranium is relatively abundant. Moreover, the development of breeder reactors gives assurance that humanity will have an abundant supply of useful energy long after it has presumably decided that the relatively available fossil hydrocarbons have become too precious to use as fuels. Also it may eventually, in the next century, prove practical to derive useful energy from the fusion of deuterium which is available in essentially infinite abundance.

10. Exhaustion of Materials

Darwin is on safer ground in expressing concern about the exhaustion of the richer mineral deposits. This, however, poses a problem of a different kind since the elements involved are not lost through use but merely dispersed so that the issue becomes one of conservation and concentration—problems that are inherently soluble, particularly if adequate energy is available.

It should also be kept in mind that the age of synthetic polymeric materials, primarily made of relatively common elements such as oxygen, hydrogen, carbon and nitrogen, had barely dawned at the time Darwin's book was written. Experience gained during the past forty years shows quite clearly that such substances can often replace materials using less abundant elements, as well as for natural products that are expensive to produce. In fact, the synthetic materials are already indispensable for the fabrication of many articles in common everyday use and the trend is growing. In a similar way we may note that components made of silicon, quartz fibers and the like, the elements of which are not very rare, play increasingly important roles in electronics, communications, and information processing systems.

11. Infectious Diseases

Darwin's expectation that infectious diseases, which once were a major factor in controlling population size, were soon to be eliminated was premature. Not only do some of the most virulent bacteria become immune to commonly used antibiotics, partly through misuse of the latter, but many types of viruses remain beyond the reach of present-day medical technology. The ongoing struggle with the virus responsible for acquired immune deficiency syndrome (AIDS) gives some indication of the potential challenges such infectious agents can offer. Had the AIDS virus been more generally transmissible through an insect vector such as common mosquitos, or in a simple way through the respiratory system, humanity might well have faced a situation somewhat analogous to that which occurred during the Black Plague which first struck Europe in the 14th century and killed of the order of a third of the population. That plague, incidentally, occurred at a time when the European population had grown to the very limits of its previously abundant food supply. Disease and famine become close partners. Along the same line, it is estimated that the dense populations of South American Indians that came in contact with the early European explorers were cut to substantially less than half—some say to a tenth—as a result of exposure to infectious agents which in Europe were responsible for so-called common childhood diseases.

In brief, a large, dense, and impoverished population such as one finds in some urban centers in Asia and Latin America, and which does not have immediate access to all benefits of forefront medical research, is a likely target for devastation as a result of a pandemic sooner or later. In the age of science-based technology such a population group is at a disadvantage relative to one at the forefront of our scientific civilization. There will be ever greater pressure with the passage of time for populations to control their numbers in order to enjoy a better life, granting that some very radical changes in attitudes and traditions may be necessary in many cases before such response occurs.

12. Intelligence

Darwin's concern with respect to the lowering of intelligence in the technically most advanced countries as a result of the low birth rate among the members of the upper brackets of society does not take into account the fact that most advanced societies benefit from the immigration of highly able individuals from less favored population groups in which the more intelligent and capable have not necessarily had low birth rates. While there seems to be little doubt that genetic composition plays a significant role in determining the level of intelligence of the individual, we know far too little about the factors affecting it to believe that mankind as a whole is suffering from a loss of the gifted in the most fundamental sense at the present time.

In addition we do not know for certain to what extent the generation and use of science-based technology will become accepted throughout the world. The obvious advantages suggest that sooner or later it could become essentially universal. It does not follow that in all such societies the upper segments of the populations will have a preferentially lower birthrate than other segments. One suspects that the birthrate at the top of Japanese, Korean and Taiwanese societies, which are increasingly dependent on the use of science-based technology (currently generated elsewhere for the most part), is at least close to the average for the entire society at the present time.

13. Importance of Fluctuations

If there is a single major flaw in Darwin's thesis, I believe it lies in what might be called insufficient recognition of the fact that the human species has occasionally demonstrated that it is capable of generating major fluctuations in both its psychological outlook and its material well-being. As mentioned earlier, the first three stages of transition in civilization noted by Darwin were probably made under duress. The fourth, namely, that associated with the scientific revolution, was partly accidental and adopted by choice.

Transitions such as these played a major role in Toynbee's account of crises that have led societies either to stagnation or a new higher level of success. It is as if humanity could, in times of great stress and crisis and with adequate resolve, call forth something in the nature of Maxwell's demon to forestall what might seem to be an inevitable fate. Since Darwin admired Toynbee's work so much, it is remarkable that he did not make more of such extraordinary responses to the challenges of a crisis in as much as they offer the possibility of hope for humanity along the pathway that lies ahead. Indeed, a number of events which have occurred in the past forty years seem to offer something in the nature of a thread of hope. Let us explore this issue further.

14. Conditions for Long-Range Survival

The guarantee of perpetual peace is nothing less than that great artist, nature (natura daedala rerum). In her mechanical course we see that her aim is to produce harmony among men, against their will and indeed through their discord. As a necessity working according to laws we do not know we call it destiny. But, considering its design in world history, we call it "providence," inasmuch as we discern in it the profound wisdom of a higher cause which predetermines the course of nature and directs it to the objective final end of the human race.

Immanuel Kant's essay On Perpetual Peace (1795), translation by Lewis White Beck (The Liberal Arts Press, 1957).

While natural genetic evolution moves at such a slow pace that one may expect changes only over a period of many millennia, social evolution can occur far more rapidly. Herein may well lie what hope we have for the future. It is conspicuous, for example, that during the last forty years the countries of Western Europe now view one another in a manner completely different from the way they did earlier in the century. Moreover, we have hope that the change is a permanent one. In brief, it appears that those nations are belatedly adopting something in the nature of a creed—to use Darwin's terminology—which would make warfare between them of the type we have witnessed earlier highly unlikely. Recognition seems to be dawning that in the scientific age cooperation is far more rewarding than the benefits of a military victory.

This view is reinforced by the fact that modern technology offers wealth and well-being to all who use it appropriately. Such transitions are not new. In the 16th and 17th centuries Europe was the scene of viciously destructive religious strife. This ended when the concept that the European people could live and indeed prosper with more than one form of religion finally dawned and was built into creed. The Peace of 1648 did not end petty bickering and dissention with respect to religious matters but it did end highly destructive and otherwise wasteful conflict. Moreover, as a byproduct the new outlook freed the Europeans to press on with the great scientific revolution which followed.

Perhaps it should be added that it seems beyond imagination that the United States would today become involved in a civil war even remotely resembling that which took place in the last century and caused so much death and destruction.

One major requirement for the long-range survival of humanity in what might be called a livable state would seem to be an essentially universal worldwide agreement at the level of a creed—an ecumenical agreement if you will—that makes warfare seem unreasonable as a solution to problems. This does not mean that human affairs will ever be devoid of everyday conflicts between individuals or groups of the type that are commonplace everywhere. Massive, destructive general warfare must, however, come to be regarded as entirely out of bounds in much the same way that internal religious warfare is unthinkable in China. Any violence that exists must be reduced to a low, possibly sporadic level with relatively little loss of life or property. International organizations and local governments will, however, need to deal with everyday conflicts between individuals and groups indefinitely, humans being what they are.

It is important that this state of affairs be achieved not by the enforced actions of a globally powerful, centralized government that could not be sustained indefinitely, but rather that it arise through the will of the populations as they seek a more rational way to live within their resources. That such a state of affairs could conceivably take place may well have seemed absurd to Darwin in the immediate wake of World War II. A half century later one is permitted at least an infinitesimal ray of hope, given time and good luck.

It is hardly necessary to add that whatever governments exist in this to-be-hoped-for state of affairs must be open in the sense that one group should possess no basis for having serious doubts with respect to the intentions of its neighbors. It follows that the governments must be freely elected and sustainable.

It is sometimes said that the existence of a universal pattern of democratic societies would alone guarantee universal peace. This thesis is questionable. The fact that the Western European governments had democratically elected parliaments in 1914 did not prevent World War I. In fact, the parliaments voted for war. The will to abolish war must lie at a far deeper level in the human psyche.

15. Population

As we have seen, one of Darwin's principal concerns was a relentless growth of population which would eventually stretch the ever-dwindling natural resources of the planet beyond the possibility of providing a reasonably civilized life. It does not seem possible at present to say what the rational limit to the human population should be. Quite prosperous modern countries have relatively high densities of population—Holland being an example. However, it does seem safe to say that it is important that the population be under control, educated to its capacity and that the societies maintain a reasonably uniform level of technical development in matters such as agriculture, medicine, communication and transportation if we are to achieve long-range survivability. As emphasized earlier, an undernourished, illiterate population which does not have the capacity to deal with natural and manmade crises is not at an advantage relative to a highly developed and educated one in spite of its numbers.

Is there indeed any hope that the world population will stabilize of its own volition in the near future and perhaps even contract in regions in Asia and Latin America in which poverty reigns supreme for a majority of the population? What hope there is, albeit a slender one, seems to lie in the continuing advance of general education which is encircling the globe with varying speed in different quarters and which leaves in its wake many benefits, including realization at the individual level that there are advantages to limiting the number of one's progeny if they are to have a rewarding and productive life. In these days of relatively low infant mortality a very large family is a disadvantage rather than an asset for most individuals.

There seem to be three major reasons the people in poor countries have large numbers of offspring. Probably the most important is to assure a source of support in old age in cases in which other measures are unreliable or even not available. Second, is gross ignorance of, or religious opposition to, the means of birth control. Third, is a psychological factor wherein the male parent gauges his own virility to a great extent in terms of the number of children he has fathered. This seems to be a major factor in some Latin American populations, and indeed in some American ghettos.

Generally speaking, societies which have accepted modern education, developed modern industry and have access to modern medicine have tended to control their populations within acceptable limits and, in consequence, have raised the standard of living and health of their citizens. A typical example is provided by Taiwan which emerged from World War II with a population of nine million persons living more or less at a subsistence level. Today it is a thoroughly modern, industrialized country. The population grew very rapidly at first and has since climbed to twenty million. However, the birthrate has fallen greatly as the benefits of having a small number of well-educated progeny in a socially responsible society have come to be appreciated. In fact, the population growth has dropped from a high above three percent to about one percent in forty years. This example illustrates the great importance of the role which the educated upper classes can play in advancing the well-being of a population. The so-called Taiwan miracle would not have occurred if the leadership had not placed the matter of advancing the status of the entire population as a top priority. It remains to be seen if and when similar policies will be adopted by the leadership in other countries which face problems much like those Taiwan did initially.

While the possibility that groups with large, impoverished populations could overwhelm the more advanced ones which have a policy of controlling their growth seems more and more remote, it appears that in the long run it is necessary that there be universal acceptance of population control. Such a condition would seem to be necessary before we can hope to attain a state in which the very concept of large-scale warfare is made obsolete.

16. Religion and Other Creeds—A Controversial Issue

Darwin discussed the importance of creeds in human affairs. However, when considering the survivability of civilization, he apparently did not give them the same status, in worst cases, as issues related to over-population and the exhaustion of readily available natural resources. He did note that some individuals are willing to face martyrdom for beliefs and that many more are ready to commit murder in support of a creed. He seemed, however, to regard such hazards to be of a lower order. On the other hand, it is quite possible that he was sufficiently pessimistic about the future of humanity for other reasons that he felt it desirable to avoid the problems that could arise from introducing emotion-engendering issues related to religious beliefs or their equivalent.

Many individuals are able to face the process of living as part of the human family in this vast universe without feeling the need to speculate about matters beyond their everyday affairs, such as those related to family, health and the means of acquiring a livelihood. Most people, however, have the additional need to be guided by a pattern of beliefs and opinions that in one way or another transcend their everyday personal affairs—they need a religion or a set of creeds to provide a broader and possibly more profound platform of support to give meaning to their lives. Experience shows that the range of such beliefs is well nigh infinite. Some creeds may be highly personal, directed almost entirely to the inner world of the individual. Others are designed to satisfy the requirements of those who find added sustenance through communion within a well-organized group. Some creeds or religions may be of a highly esoteric and spiritual nature. Others, as has been true of national socialism and communism, may deal with far more earthly matters related to ethnic, national or economic affairs. Most contain an admixture of beliefs related to both heaven and earth. Some creeds are entirely benign in the sense that they have a bearing only on the practioners. Others may reach far beyond that group and produce disturbances, including violent conflict, over significant portions of the human scene. It seems that creeds of this more violent form deserve consideration at the same level as matters related to exhaustion of resources and over-population.

In our own day, both national socialism and communism have engendered bloody, destructive conflict on a global scale. Only a short while ago, in historical terms, the Inquisition did its best to prevent by force the advent of the scientific revolution and the science-based form of civilization which is penetrating the world at the present time. On a lesser but by no means trivial scale, religious fundamentalists in Iran succeeded in overturning the well-intentioned plans of the previous government to introduce the methods of western science into that country as mentioned in the previous chapter. Much of the bloodshed in the Near East at present is based on a conflict of creeds that clearly has world-wide repercussions and, in the worst case, could lead to extended, bitter warfare.

Again, it seems that the danger to the continued advance of human society that can result from the activity of the more militant forms of creeds is as great as any other since they can engender so much wasteful conflict. Moreover, they can becloud the forms of rational thought, planning and action necessary in the millennia ahead if mankind is to use the opportunities that the world about us provides in a reasonably productive way.

It seems that the third condition for the long-range survival of civilization is the establishment of agreement on a world-wide scale of the principle that creeds, whether of a conventionally religious nature or otherwise, must be constrained so that they do not engender extended destructive conflict. Tolerance in such matters must become the watchword. This is a generalization of the principle that religion and politics can make very bad partners. Dealing with this issue on a worldwide scale may prove to be far more difficult than finding new energy sources, preserving precious minerals or controlling population growth—areas in which science can play an important role. Inducing changes in the thoughts and hearts of mankind can be a much more formidable task.

8
A Physicist's View of Living Systems[1]

There is an ancient nursery rhyme with which we are all familiar. It goes as follows:

Humpty Dumpty sat on a wall.
Humpty Dumpty had a great fall.
All the king's horses and all the king's men
Cannot put Humpty Dumpty together again.

Humpty Dumpty is, of course, an ordinary hen's egg, consisting of a single-celled organism linked to a highly nutrient medium and encased in a thin shell. We may well wonder if modern biologists, with all the miracles they are producing, could be more effective than the king's staff was expected to be at the time this nursery rhyme was written.

1. Advances in Biology

The field of biology has made enormous strides in recent decades upon a solid base of ingenious organic chemistry aided by instrumentation from the physical sciences. There have been many important steps since the synthesis of the first organic compounds early in the last century. Six of the most important in recent decades are as follows:

1) The utilization of the electron microscope to get beyond the very limited resolution of cellular structure provided by the optical microscope. Indeed, the electron microscope permits resolution essentially down to molecular dimensions.

2) The use of the high-speed centrifuge to separate cell constituents such as organelles for analysis following disruption of the cell membrane.

---

[1] A version of this paper appeared in the *International Journal on the Unity of Sciences* (Volume 2, Number 3, p. 349).

3) The discovery that DNA carries the genetic message. This discovery was first demonstrated with a type of bacterium responsible for pneumonia but then found to be general.

4) The discovery that DNA is double stranded and linear in diploid cells.

5) The use of special isotopes of the elements as tracers in chemical reactions.

6) The development of computer-aided techniques based on x-ray, neutron diffraction, and nuclear magnetic resonance for the determination of the arrangement of atoms in molecular structures. This has permitted almost routine determination of the structures of many biological molecules once even relatively tiny crystals become available.

It is obviously possible to add additional major landmarks such as the discovery of the way in which the DNA in genes serves as a template for the generation of proteins, the solving of the genetic code which determines the ways in which the four codons of DNA translate into the twenty amino acids which appear in proteins, and the discovery of cell membrane proteins as well as of many other details of cell structure and mechanics.

2. Eukaryotic and Prokaryotic Cells

It is well known that existing self-reproducing biological cells come in two forms, namely, eukaryotic cells, which contain a nucleus in which most of the genetic material resides in a highly structured way, and prokaryotic cells, an early evolutionary form in which the genetic material is outside the nucleus and can be relatively free. Most multi-celled organisms, with the exception of some primitive plants, have eukaryotic cells. Bacteria are prokaryotes. At the present time some investigators are making comparisons of the common elements in the genetic material in different types of prokaryotes to see if they can gain information concerning presumably extinct ancestral progenitors. While those ancestors probably were simpler than present day bacteria, they undoubtedly were fairly complex mono-celled organisms possessing most of the rudiments of the structure of the prokaryotes.

It should be added that fossilized impressions of primitive cells have been found in geological deposits about 3.5 billion years old. Such dates are to be compared with the age of the earth which is estimated to be about 4.5 billion years. Living systems appeared relatively early in our planet's history under conditions which are beyond our present understanding.

A typical eukaryotic cell such as we all have in our bodies, has overall dimensions on the order of some microns (0.0001 cm) depending on the type of cell, so that it contains in the range of 10^{12} atoms. Moreover, it possesses a complex structure. This includes the nucleus, in which the genetic material is located along with a protein backbone, and a complex mechanism termed the endoplasmic reticulum for translating the information contained in the nucleus into proteins which can serve in multiple ways. It also possesses tubules which provide a skeletal structure and serve as guiding conduits for transportation of signals and cellular material as well as cell surface proteins which monitor transport into and out of the cell, serve as receptors for special molecules in the ambient, and play many other roles. Moreover it contains a chemical powerhouse, termed the mitochondrion, and in some cases a packaging device termed the Golgi apparatus which can produce storage units containing such items as hormones or digestive enzymes that can be sent out of the cell to serve other parts of the organism.

In the case of plants, the cell would also contain a photosynthetic unit or organelle for converting solar energy into storable, energy-rich compounds. One could provide more details but this rudimentary discussion will suggest the complexity of the cell and its functions. To survive, it must, of course, be embedded in some form of nutrient aqueous medium.

3. Viruses

For completeness, it should be mentioned that viruses, which tend to permeate the biological world in a parasitic way, are simple cells structurally but cannot reproduce themselves without the aid of more complex cells which have the ability to reproduce. They invade such cells and take control of at least part of the reproductive system.

4. Reductionism

The recent brilliant successes achieved in the field of cellular and molecular biology have given additional convincing support, if any was needed, to the view that the essential molecular building blocks of biological systems obey the normal laws of physics—including quantum mechanics—and encourage the opinion that the operation of living systems is essentially mechanistic, in a sense similar to, although more complex than, a man-made machine. This view, which is indeed prevalent at least implicitly in the biological community today, is designated by the term reductionism wherein complete knowledge of the whole system is expected to be gained by understanding the nature of its parts.

The notion of reductionism is an ancient one, fundamental to the art and science of technology. One of the best ways of seeing how a mechanical, optical, or electrical contrivance works or fails to work is to take it apart to see how the parts interact with one another—a process known as reverse engineering. Moreover, the design of new equipment often starts with basic notions concerning components which are then modeled in one way or another to fit together to achieve the desired working system. True reductionism implies something in the nature of a two-way street. While the whole may be regarded as more than the sum of its parts in such a physical system, its action can ultimately be understood in terms of those parts and their interplay. Can this two-way concept be applied at the present level of our understanding of biological systems?

Perhaps it should be added that one of the interesting exercises carried out by photochemists, which is often quoted as a triumph of its kind (and with which one cannot disagree), is that some of the building blocks of biological systems, such as amino acids, can be generated from inorganic constituents by subjecting a gas or liquid containing the latter to ionizing radiation. To those who take the most extreme reductionist viewpoint, this is regarded as almost a closing link in supporting the view that modern molecular biology has truly reduced the fundamentals of biology to those of everyday physics.

It should be emphasized at the start of an examination of the role of reductionism in biology that there is no reason to doubt that biological molecules, as such, obey the same physical laws as inorganic molecules. However, we must look into the question of the degree to which understanding the laws governing the properties of the molecules of which living systems are composed permits us to understand the properties of a living biological system such as the cell. To do this, let us attempt to follow the logic and procedures used in discussions of many-body systems commonly considered by physicists and chemists in treating inorganic systems such as atoms or molecules, or condensed matter such as crystals, liquids and the like.

5. Conventional Theoretical Approach

Physical chemists have done much to apply quantum mechanics to the understanding of some properties of complex molecules, following earlier successful work by physicists in relation to simple atoms and molecules composed of the lighter elements. Some of the work of such chemists contains a degree of arbitrariness as a result of the use of quasi-empirical parameters, but there is no doubt that one can obtain valuable knowledge concerning the structure and other interesting properties of the individual molecules, including aspects of their interactions with other molecules using such techniques. Indeed, modern computers, when combined with sophisticated picture tubes, have made it possible to visualize three-dimensional aspects of such modeling in a most remarkable way, including visualization of simple dynamical features of the systems.

Since large molecules behave to a degree like assemblies of somewhat independent component groups or radicals, the techniques can probably be carried very far as computers become more and more powerful and flexible. It remains to be seen, however, if limitations will arise from the fact that the three-dimensional or stereo aspects of biological molecules, including the presence of cavities or projections, must be visualized as well as the gross structure. Such specialized features, particularly the surface features, can determine essential characteristics of large convoluted biological molecules.

It might be added parenthetically that some of the greatest quantitative successes in the application of quantum mechanics to multi-atomic systems have been achieved in working with inorganic single crystals. Although crystals may contain large numbers of atoms, they are basically simple in nature because of their lattice structure. The same grouping of atoms is repeated endlessly in three dimensions so that they actually do not have the same type of complexity as large biological molecules.

Granting that the biochemists have provided us with an enormous amount of useful information regarding the detailed structure of a living cell, we may now properly ask if we are able to use this information in a reasonable way to obtain a holistic picture of the workings of a cell comparable to what we normally expect for physical systems. To be realistic, we should consider in principle at least an ensemble of about 10^{12} atoms containing in the simplest case at least several hundred different kinds of molecules, each with a degree of autonomy not found in simple crystals.

In fact, the biochemists tell us that a typical eukaryotic cell, such as one in our body, contains at least 1,000 different types of protein molecules alone. The reader should understand that no one would actually try to program such a problem from the basic equations of atomic physics in practice, but it is important to understand what would be involved if the attempt were made.

6. "Immensity" of Problem

The distinguished physicist Walter Elsasser first pointed out, about thirty years ago, that for a living system, the inverse problem of constructing the whole from its parts in the abstract from first physical principles is hopelessly difficult from the human viewpoint because of what he terms the immense complexity involved. This is true regardless of what types of convenient shortcuts or approaches we may be willing to make in the process. For one thing, the number of possible arrangements of atoms or molecules in the cell which should be compared with one another in the search for the living arrangements, is enormously large. To obtain perspective, we may recall that the number of permutations and combinations of a simple linear array of n objects is n! (factorial) which is approximately n^n. Thus, if n were somewhat arbitrarily chosen to be 100, which is a tiny number by our standards, the number of possible arrangements would be approximately 100^{100}—a simply colossal (immense in Elsasser's terminology) number from the human standpoint, namely, 10 followed by two hundred zeros. It is enormously greater than the number of fundamental particles in the universe. Actually, the complexity of the problem of trying to assemble a biological system from its parts is immensely greater since in practice we have many more parameters than merely 100.

Another point should be made. Of the immensely large number of ways of arranging the atoms in a cell, regarded as a physical ensemble of atoms or molecules, it is evident that only an immensely small fraction can represent viable living systems. The vast majority of arrangements correspond to nonliving states of the aggregate. The probability that we could accidentally assemble a living configuration from inert ingredients is immensely improbable.

The computer theorists have a term, "transcalculational," for problems too complicated to solve exactly by any imaginable device designed by human beings.

The problem of modeling a standard chess game, which involves only 24 pieces, from start to finish, considering all alternative moves, is transcalculation. The computers which play chess—and they are very successful at it—analyze the possibilities in a given situation only a few steps ahead, not for an entire game. The problem of putting the mechanics of the atoms or molecules in a cell on a computer is immeasurably more complex and is transcalculational at a much higher level.

7. Statistical Approach

It should be mentioned that in some games analytical procedures which determine statistical probabilities or "odds" associated with individual decisions can be of great value to the players even though an exact analysis of a complete game may be transcalculational. In fact, statistical procedures are also commonly used in coming to grips with many physical problems involving stochastic processes, such as turbulence in fluids. They do not appear to have any major place of a comparable kind in dealing with living systems at the holistic level for two reasons. First, living systems possess a vastly larger number of parameters. Second, the essence of the operation of a living system is a high degree of regulation and order, or integrated design, rather than the type of chaotic organization to which statistical methods apply.

One might add, however, that many scientists believe that such design has been achieved from the start as a result of purely accidental interactions among fairly ordinary molecules, conditioned by natural selection of the fittest. Indeed this belief has achieved the status of an item of faith or dogma for many in the profession. Actually it should be regarded as a major unsubstantiated assumption, still open for scientific inquiry, granting that the matter of the evolution in overall complexity of ongoing biological systems over time is beyond reasonable doubt.

What the modern biologist does have as a powerful tool is the ability to explore, often with exquisite skill and in great detail, a microscopic segment of the whole living system. This permits the profession to gain useful information and what might be called peripheral insight into what is the essentially miraculous working system of a living cell. While this process may yield a finer understanding of some selected family of components of the entire system, it does not by any means provide holistic understanding of the immensely complex system. That appears to be well beyond our reach.

8. Direct use of Physical Equipment

To elaborate the basic theme further, we may ask about the possibility of using imaginable physical equipment to observe the overall details of the working of a typical cell at the atomic or molecular level. Quite apart from the unmanageable complexity of keeping track of a vast number of different molecules simultaneously, one runs into a problem cited by Niels Bohr more than fifty years ago, soon after the development of wave mechanics. He noted that physical measurements carried out on a living cell with sufficient precision to determine atomic positions, for example, with the use of an x-ray microscope having resolution at the atomic scale, would be certain to kill the cell and hence defeat the very purpose of the experiment. He regarded this as an aspect of the principle of complementarity: One can have a living cell without knowing all the facts about its internal dynamics, or one can have considerable knowledge about the internal dynamics of a dead cell. While we have learned much about the internal constitution of cells since Bohr's statement, the basis for his observation is still valid.

9. Problem of Individuality

Elsasser has pointed out that in dealing with living systems—for example, single-celled organisms—we do not have the great advantage that we possess in analyzing the properties of individual atomic or molecular species. We can, in principle at least, regenerate the state of any type of molecule—biomolecule if you wish—indefinitely and carry out an unlimited number of identical experiments upon it in sequence. Or, to use the standard terminology of quantum mechanics, we can always regenerate the same initial state indefinitely, at least in principle, as is done in molecular beam experiments in which one has a continuous stream of atoms or molecules in the same initial state. Even the simplest line of living systems exhibits what might be termed individuality, so that we cannot hope to reproduce the same original state in the sense of the physicist for identical repetitive observations. In general, clones of a given cell differ in essential ways from their parents. This represents one of a number of fundamental differences between living systems and merely complex, but reproducible, physical systems.

10. Large-Scale Modeling

One may emphasize the difference between what might be called normally complex and immensely complex systems in the following way. We all know that any large-scale geophysical system such as the atmosphere, the oceans, or the convecting matter of the earth's interior is complex. Such systems possess, among other things, stochastic features which may make precise solutions outside the reach of normal humans except in the crudest statistical sense. Nevertheless, one can imagine, in principle at least, a superhuman intelligent being of galactic dimensions, in the manner of some of Fred Hoyle's science fiction creations, who could replicate to any degree it wishes an indefinite number of copies of the earth from galactic material, and by such replication, develop an understanding of the basic geophysical dynamics. The being could recreate a variety of initial physical conditions for the earth and obtain a very good picture, at least in a statistical sense, of any important aspect of its behavior. A comparable attempt to understand the dynamics of even the simplest living system would be completely beyond the reach of such a being because of the higher degree of what we have termed the immense complexity of the living system.

It should be added parenthetically that modeling the geodynamics of our own planet may not provide the best example to consider for this purpose. There are individuals who suggest that the presence of life on Earth has had a profound influence on its geophysical properties. This proposal is sometimes known as the Gaia hypothesis. Perhaps the arguments just made would be better applied to planets, such as Mercury, Venus and Mars, which presumably have not been affected by living systems to the best of our knowledge.

11. Molecular Origins of Life

Early in this discussion mention was made of experiments carried out using radiochemistry techniques which demonstrate that, among other things, amino acids can be produced when the constituent elements are subject to ionizing radiation. While one can only admire the achievements of this work, anyone who has long had an involvement in the effects of ionizing radiation upon materials will point out that it would indeed be surprising if traces of almost any reasonable combination of the elements in the system did not appear in the experiments. What is more, one must admit that it requires a long leap of imagination and much hand-waving to fill the gap between the creation of such elemental organic molecules by ionizing radiation and the development of the living cell. We would not expect to bridge this gap by stirring the raw molecular ingredients found in the living cell in a test tube.

Several years ago the author was part of an audience that had the privilege of hearing Freeman Dyson give an informal talk on the possible way in which the first truly metabolically stable entity might have emerged from elementary material. His presentation eventually became part of a small book.[2] His starting "physical unit" involved some ten thousand atoms which is of course much too complex to model from first principles in any precise way. His discussion was, of necessity, qualitative and symbolic. Most of those present were biologists who were taken aback by the fact that Dyson did not place primary focus upon the generation of DNA or RNA but rather on the issue of metabolic stability by whatever route it may have come about. This dismay says a great deal about the degree to which the modern biologists have placed the holistic problems of their field in the background in their eagerness to get on with the admittedly important process of analyzing the properties of the molecular constituents of existing types of cells.

---

[2] F. Dyson, *The Origins of Life* (Cambridge University Press, 1985).

12. Qualitative Difference

Where does this leave us in thinking about the physical aspects of living system? We must first recognize that the concept of reductionism as applied to biology must be regarded as qualitatively different from that applied to inorganic, physical systems. We can gain an essentially unlimited amount of information concerning the constituent molecules of living systems, but are vastly limited in gaining understanding of their detailed workings, except in a very piecemeal and unintegrated way. Whereas the physical scientist usually faces only what might be termed earthly limits in dealing with his or her domain—available budgets and limits of time and size—the biologist has a far more cosmic limitation of a fundamental kind, somewhat analogous to the limits expressed by the Heisenberg uncertainty relation. Indeed, essential details of matters, such as the processes of biological evolution, may be obscured from us because of such limitations.

13. Elsasser's Views

Concepts of this type are very eloquently expressed by Elsasser in a recent book, *Reflections on a Theory of Organisms*.[3] Central to his viewpoint is the opinion that the state of physical complexity of living systems, with all its implications of individuality and reproducibility, is not only characteristic of, but is basic to, the existence of such systems. The relationship is an entirely "natural" one and one of the potential attributes of immensely complex systems. The understanding of such a system in a completely holistic sense is, however, not possible for us using the techniques normally applied when analyzing physical systems.

Elsasser associates two important attributes to complex physical systems of the appropriate kind. First is an attribute designated by him as "creativity" with the potentiality of generating a living system, in spite of the apparent improbability of the process when judged from the standpoint of everyday statistical physics.

Second, such a living system is perforce endowed with what Elsasser has designated "biological memory," that is, the ability of the system to sustain internal order through many biological generations, in spite of the natural tendency of physical systems to move spontaneously to states of ever higher disorder. The entropy of the environment in which the living system resides may increase as a result of cell metabolism, but the disorder in the living cell does not. It appears that the understanding of these attributes is probably outside the reach of the approaches commonly used in the physical sciences. They are inherent to appropriate types of complex aggregates of molecules and, above all, to living systems.

---

[3] *Reflections on a Theory of Organisms* (Orbis Publishing, Frelighsburg, Quebec, Canada, 1987).

Please understand that there is no attempt here or in Elsasser's work to offer any mystical viewpoint of the issues, but only a desire to express the sense of limitation one experiences when trying to employ the reasoning and the techniques of conventional physical science to biological systems. We deal not with the occult, but with a major limitation on the use of a common type of physical analysis when attempting to apply it to biological systems. Indeed, we appear to deal with what might be termed the "uncanny" in the original meaning of the word—the "unknowable." To quote Elsasser's final paragraph

> In conclusion, biology is the realm of natural phenomena that is dominated by the concept of creativity, whereas physics is, and always has been, a deterministic science. This distinction, together with a new kind of phenomenon in organisms, namely, holistic memory, constitutes the main results of this inquiry. Furthermore, we have formalized the well-known fact that the synthesis of life *in vitro* encounters insuperable difficulties.

None of this discourse implies that we should discourage the molecular biologists from carrying on their present forms of research and associated analysis. Such work not only provides important enlightenment concerning detailed aspects of living systems, but will, inevitably, create a revolution in medical science, piecemeal though it may be. One can only hope, however, that the profession eventually becomes aware of the fact that speculations concerning the holistic aspects of biological systems may be as worthy of attention as the details to which so much attention is now given.

Epilogue

Let us close by asking whether we have been grossly unfair, in some major sense, to the spirit of modern biological science in the foregoing presentation. The answer is probably mixed. To the physicist who is concerned with understanding the basic laws which govern the phenomena we observe in the world about us, it is not possible to ignore the special attributes of living systems when looked upon in a holistic, as well as in a detailed, manner. We here attempt to underscore what, for the present at least, can be termed the enigmatic (but natural) features of living systems that must continue to be the focus of serious attention whatever conclusions may ultimately be reached.

The biologist, on the other hand, must inevitably do what can be done to understand detailed features of individual biological systems and usually finds it distracting to be called upon to devote attention to what might be termed larger issues. This is apt to be particularly true at a time, such as the present, when the frontiers associated with molecular and cellular biology are in such a dynamically productive state. Perhaps the very success of the biologists in dissecting cells into their constituent organelles and molecules, along with the ability to synthesize some of the latter, has overcome their rightful sense of awe at the existence of living matter. The physicists, who have labored in their own way to reveal in part the laws governing the cosmos, the atom and the subatom, have had their sense of wonderment at the grandeur of nature enhanced. One can only hope that a time will come when the interests of the physicists and the biologists will become more closely fused.

9
Nuclear Science:
Promises and Perceptions[1]

The discovery of radioactivity at the turn of the last century and the discovery of fission at the end of the 1930's was accompanied in both cases by the emergence of socio-political activism on the part of a number of prescient individuals.

Some of the response was positive as well as prescient in the sense that individuals saw on the horizon the potential for an unlimited source of energy for the extension of industrial civilization on a world-wide basis. Others were concerned mainly with negative aspects. Indeed some individuals have, for their own reasons, sought to generate public fear without offering any compensating form of balance or enlightenment. This bifurcation of outlook is not new in human history. The age of steam, the rise of chemical technology, the dawn of the Air and Space Ages and the great discoveries in the field of molecular biology have produced similar divided emotions and activities. Every emerging technology seems to generate a mixture of hope and fear.

Much of the history of this development is contained in Spencer Weart's book *Nuclear Fear*,[2] but some personal touches will be added. In particular, we shall examine aspects of the activism by focusing on the roles played by two excellent scientists who were prescient and motivated primarily on the basis of humanitarian concerns.

The first is Frederick Soddy whose life extended from 1877 to 1956. He is scarcely remembered at the present time although he was clearly the first person

---

[1] This chapter is based on a presentation delivered at George Washington University in Washington, D.C., in January of 1989 on the 50th Anniversary of Niels Bohr's announcement at a meeting held at that university of the discovery of nuclear fission. The author was present at the original meeting in 1939.

[2] Spencer R. Weart, *Nuclear Fear—A History of Images* (Harvard University Press, 1988).

to appreciate the long-range potentialities associated with the discovery of transmutation of the elements. The second is Leo Szilard, the brilliant Hungarian-born physicist who appeared in the world in 1898 and died in 1964.

In the areas of the physical science, we can apparently recognize two concepts that seem to be imbedded in the human subconscious and which surface in one form or another in each generation among individuals who are appropriately stimulated. One is the concept of the frequent visitation of beings from a different world, terrestrial or otherwise, who permit themselves to be seen only occasionally. Beings such as leprechauns and trolls held the field in the past. In our own time this phenomenon appears in the form of observations of unidentified foreign objects or UFOs. There was a rash of claims of such visitations in the 1950's and 1960's and a new one in the 1980's.

The other concept is related to the possibility of the development of an agent or instrument that can threaten the survival of a large part or all of humanity and can lead either to our destruction or our salvation. This concept appears in various forms throughout recorded history. In the last century it emerged in the literature in the form of weapons made possible by discoveries which take place at the frontiers of science. It is, for example, a theme found in several forms in the clairvoyant books of Jules Verne—an author with whom Soddy was familiar.

The author first encountered this form of psychic experience at a relatively primitive level when serving in a technical intelligence office in General Eisenhower's headquarters in Europe near the end of World War II. The office was headed by Dr. H. P. Robertson, the expert on relativity theory who had been at Princeton University in the 1930's. A German private soldier who obviously suffered from a severe psychic disturbance came to the attention of our military staff. He was interviewed and a detailed report was sent to our office. He believed that he had witnessed, under unlikely circumstances, the test of a bomb by his own general staff and scientists that was many orders of magnitude larger than anything then available anywhere and which destroyed an area of many square kilometers. He described the incident in great detail. This was well before the July 16th test at Alamogordo.

1. The Start of Nuclear Science

In 1896, Henri Becquerel discovered by accident that a specimen of uranium-containing mineral in his laboratory emitted a penetrating radiation which could expose photographic plates in sealed packages. This was followed soon after by the isolation of radium—a minor constituent in pitchblende ore—which on a weight basis was an even more intense emitter of similar radiation. The natures and ultimate origin of the radiations were completely unknown and became the object of a great deal of fanciful speculation. Coming as they did at the same time as the experimental isolation of the electron as a constituent of the atom and the discovery of x-rays, the new disclosures added immensely to the excitement current in the field of physics as it entered into a turbulent, revolutionary period.

By the summer of 1903, Ernest Rutherford and Frederick Soddy, working in close collaboration at McGill University in Montreal, had succeeded in demonstrating beyond serious doubt that the phenomenon of radioactivity, the term used to designate the effects found in uranium, radium and a few other heavy elements including thorium, was intrinsic to the atoms of the species which exhibited it and involved a transmutation of elements. Soon thereafter, in 1905, Egon von Schweidler, a colleague of Ludwig Boltzmann, introduced the present-day concept of the statistical nature of the disintegration process.

Incidentally, the somewhat accidental partnership of Rutherford and Soddy was a most remarkable happenstance. Rutherford, a few years older than Soddy, was a reasonably well-established experimental physicist and Soddy was a newly graduated physical chemist with a strong interest in the history of chemistry. The experience of both was essential to the discovery that radioactivity arose from atomic disintegration.

There are two important social issues associated with these developments which should be mentioned. The first is related to professional attitudes. One might have supposed that the older well-established chemists would have been especially intrigued by the discovery of radioactivity and plunged into the field. The opposite seems to have been the case. It is almost as if they hoped that the observations would go away. One suspects the issue was partly psychological in the sense that they were involved in other successful work and partly related to the fact that, as well-established chemists, they saw a threat to the hard-won concepts of the immutability of matter. Fortunately the field did prove creatively exciting to some of the younger chemists. Two young chemists deserve special mention.

First there is Frederick Soddy mentioned above. He arrived at McGill University from England in 1900 as a very junior faculty member at the age of 23. He became fascinated with Rutherford's primitive attempts to unravel the fundamentals of radioactivity and joined him as a fully dedicated partner. Rutherford, then 29, had arrived at McGill two years earlier and had begun focusing on the observations related to the radioactivity of thorium. Incidentally, Rutherford openly stated in later years that the cooperation of Soddy was essential to the success of the early work. Soddy did not share in Rutherford's Nobel Prize but was rewarded later for his contribution to the discovery of isotopes, chemically identical atoms with different masses.

The other notable young chemist who decided to devote his career to the field of radioactivity was Otto Hahn who, after completing his doctoral work at Munich in 1901 at the age of 22, first joined William Ramsey, the discoverer of helium, in London and then joined Rutherford at McGill to become one of the world's foremost radiochemists. His first important work carried out in Rutherford's laboratory was to refine the analysis of the thorium chain of reactivity made by Soddy in an essential way.

Hahn's great contribution to the field, with Strassmann nearly forty years later, was, of course, to make sense out of Fermi's somewhat rudimentary and only partially correct analysis of the effects of irradiating uranium with neutrons. Fermi correctly predicted the production of transuranic elements, but missed the fission process and several other consequences of the irradiation. Hahn reported by letter to Lise Meitner the phenomenon he termed the "bursting" of uranium. She discussed it with Otto Frisch, who then reported it to Niels Bohr who brought the news to the United States in 1939 and stirred the community of nuclear physicists, not least those in United States, as never before. Doubtless, his early work on the natural radioactive elements stimulated Hahn to focus on the confusing results for uranium that emerged out of Fermi's laboratory from their run through the periodic table. He also states in his biography that Aristide von Grosse had urged him to explore the matter further.

2. Soddy's Prescience

The second important social consequence of the early years of radioactivity relates to the effect that his work with Rutherford had upon the ultimate career of Frederick Soddy, for it led him into messianic pathways and a search for a peaceful world through scientific approaches to economics and sociology. In great contrast, of course, Ernest Rutherford regarded radioactivity as primarily a laboratory phenomenon having few auxiliary applications. As late as the 1930's he made public statements to the effect that any talk of producing significantly useful energy from the nucleus on the basis of what was known then was "moonshine."

In his book, *Nuclear Fear*, Spencer Weart expresses the belief that well before his association with Rutherford, Soddy developed strongly the view that chemistry would not only make enormous advances in the future but would have the effect of liberating humanity from drudgery and provide it with unlimited capabilities, including access to essentially free power to drive the machines of the world. With this background, it was easy for him, once the atomic disintegration with its relatively enormous release of energy on the atomic scale had been revealed, to assume that in some way, through the advancement of science, mankind would learn to gain access to such energy. He, in effect, became the first scientist apostle of the nuclear age. He turned his attention to what he regarded as scientific approaches to economics and sociology and became a wide-ranging public lecturer who attracted large and often very distinguished audiences. No doubt his earlier association with Rutherford whose fame was growing continuously played an important role in the reception he received. One of the individuals who took his predictions very seriously was H. G. Wells who, as early as 1913, before World War I, foresaw the potential dangers associated with the development of what we now term nuclear weapons and wrote a book with the title *The World Set Free* in which, among other more desirable things, such weapons are used and cause great destruction.

It is noteworthy that Well's book appeared before the start of World War I and at a time when the products of advances in science and technology were generally regarded to be overwhelmingly beneficial to mankind. Soddy seems to have retained this optimism until the horrors of that war descended upon Europe. He was particularly shaken by the death at Gallipoli of the brilliant young physicist Moseley who had demonstrated the relation between the frequencies associated with x-ray spectra and atomic number. After that he became a pessimist concerning the effects of the release of nuclear energy on a practical scale which he continued to believe was imminent.

Soddy's biographer, Muriel Howorth, makes no mention whatever of his reaction to the news of the discovery of fission or to the successful development of the nuclear bombs during World War II and the postwar period. Soddy was in his sixties at that time and still very active. One can only presume that he had become mentally adjusted to these developments long before they occurred.

It is clear that the well-established scientists in Soddy's generation felt that he had wandered far afield and lost touch with his own community. One item in a biography extols his early contribution to the unraveling of the mysteries of radioactivity but states that he lost his genius in later years.

3. Leo Szilard

The next major scientific figure to give warning that the nuclear age was imminent for better or worse was Leo Szilard who became interested in the scientific, technological and social aspects of the subject in the mid-1930's and never left it until his death in 1964. He undoubtedly was influenced in part by Chadwick's discovery of the neutron[3] in 1932 and by his friend Eugene Wigner's early involvement in the theory of nuclear structures, but his mind soon soared far above all this. We apparently know of no direct connection between him and Soddy but having moved to Oxford in 1934 when Soddy was still in his prime, he could not help but know of Soddy's predictions and anxieties. We also know that he was strongly affected by H. G. Wells' book, *The World Set Free*, which he first read in German translation.

In any event, Szilard became concerned with the possibility of a neutron-induced chain reaction using one neutron-rich nucleus or another some five years before fission was discovered. The author first met him through Eugene Wigner in

---

[3] The discovery of the neutron by Chadwick in 1932 changed the course of nuclear physics radically. The discovery has an interesting background. Soon after 1911 when Rutherford revealed the existence of the atomic nucleus by bombarding gold atoms with the alpha particles (helium nucleii) from radium, speculation began as to whether the nucleus might contain uncharged particles with a mass close to that of the positively charged proton, the single particle constituting the nucleus of ordinary hydrogen. A combination of the two would account for the atomic charge and atomic mass of a typical nucleus.

Immediately after World War I Chadwick, working in Rutherford's laboratory in Cambridge, England, carried out a series of experiments in which he bombarded a number of light elements with natural alpha particles in the hope of dislodging the hypothetical neutral particle. By chance, the laboratory did not have ready access to beryllium at that time or the experiment undoubtedly would have been successful using that element. In this case, nuclear fission might well have been discovered in the 1920's and the course of history changed. Chadwick returned to the problem in the early 1930's when French investigators, working near an accelerator, mentioned the appearance of some strange radiation.

1935 or 1936 during one of his visits to the United States. His conversation focused almost entirely on two themes, the dangers Hitler posed to the Free World and the imminent feasibility of releasing nuclear energy if the right steps were taken. A year or so later, when the author was working at the General Electric Research Laboratory in Schenectady, he was doing his best to convince the head of the laboratory, Dr. William Coolidge, to undertake experiments under Szilard's direction. As far as can be recalled he focussed at that time on using the light neutron-rich nuclei in a highly compressed state as the source of energy to be derived from a neutron-induced chain reaction. In retrospect this was a very dubious proposal for an earthbound experiment at that time, although his enthusiasm was unquestioned. As might be expected, he focused on the positive aspects of achieving access to nuclear energy. He eventually became discouraged with this approach but was immediately on hand as a leader when fission was discovered a short time later. There is no doubt that Szilard played a major catalytic role both in England and the United States in pressing for work on a chain reaction. This became an all-consuming activity.

It was the author's privilege to spend two years at the University of Chicago with Szilard during World War II. At that time his mind reverberated continuously between the potential benefits and hazards of the nuclear age, tempered always by a day to day interest in the course of the war, and speculations on the progress being made by a hypothetical German version of the Manhattan District.

4. Attempts at International Control

Once it became clear in the summer of 1945 that functioning bombs existed and that the United States alone possessed them, Szilard essentially lost interest in peaceful uses of atomic energy except in a very peripheral way and focused all his attention on the matter of control. Apparently he had first accepted the thesis that our country was controlled by a few wealthy families and business people, for the most part located in New York City, a theme which President Roosevelt used in the 1930's during some of his campaign speeches. On reading items in the press regarding Beardsley Ruml, a prominent financier on Wall Street, he went to New York, introduced himself to Ruml and arranged a meeting with a number of New York businessmen and investors urging them to take steps to make certain that future control of nuclear bombs and their development be kept in safe civilian hands. His listeners heard him out with rapt attention since many of the things which he said were new and of great interest to them. When he concluded his presentation, however, they stressed the fact that the country was run from Washington and not from New York and offered to introduce him to influential senators and congressmen. He essentially took the next train to Washington and quickly gathered together a very effective lobbying group consisting of scientists and others sympathetic to his mission.

His great success was, of course, the defeat of a bill concerning the control of nuclear energy which was then passing through the Congress, which he presumed would in one way or another extend the authority of the Manhattan District in military hands. He and his colleagues played a major role in the legislation which led to the creation of the Atomic Energy Commission.

There is no evidence to indicate that Szilard was ever consulted by Bernard Baruch when the latter dealt with the United Nations in connection with the Lillienthal-Acheson Report which urged international controls. Instead, Baruch selected as his advisors individuals who came out of a more conventional stream. Nevertheless, we can assume that Szilard kept close track of the developments through the various organizations of scientists and that he placed great hopes upon the success of some form of agreement toward international control that would override narrow national interests.

When it became clear that the Soviet Union would not accept the Baruch plan, Szilard took the initiative in his own way in an attempt to gain some form of international agreement. As one of the instruments to achieve this purpose, he formed the so-called Einstein Committee on which the author served for a number of years after returning in 1947 from a year at Oak Ridge as a colleague of Eugene Wigner and as Director of the first training program on nuclear reactors. This committee was based on the ashes of an earlier committee—the Emergency Committee of Atomic Scientists—which had exhausted itself in its earlier attempts to achieve international control. While Albert Einstein was the honorary chairman of the new committee, most of the planning and action was carried out by Szilard and Harrison Brown, a geochemist who had worked with the Manhattan District and was then at the University of Chicago.

One of Szilard's greatest hopes in the period between 1946 and 1949 was to attempt to arrange a meeting of scientists from the United States, the United Kingdom and the Soviet Union at a neutral place, preferably a Caribbean island such as Jamaica or Trinidad, in order to discuss means of establishing international control over nuclear energy in general and nuclear weapons in particular. Several meetings were arranged between Harrison Brown and Soviet Foreign Secretary Andrei Gromyko to discuss this proposal, but at the end, Gromyko was compelled to say that his country opposed such a meeting. What we did not know then and which came to light in 1949 was that the Soviet Union had obtained sufficient information, presumably through Klaus Fuchs and possibly others, to get on with its own program for a fission bomb much more rapidly than most of the experts in the government had believed possible. The genie was out of the bottle. Szilard's dream of such a gathering of scientists took a full decade to be realized and is now reflected in the meetings associated with the name Pugwash.

One is inclined to believe that this failure at that time in the cycle of events, accompanied by the Soviet rejection of the Baruch proposal, had a very profound effect on Szilard and that henceforth he believed that a very destructive world war was inevitable. National ideology had gained the upper hand.

5. Aftermath

By the time the Pugwash movement had come into being, Szilard was no longer at center stage. In fact he was devoting much of his time to biological research, at which he was highly innovative, and to writing intriguing science fiction. In one of the histories of the Pugwash movement published in England, an English group is mistakenly given almost complete credit for its conception and realization. Szilard, who had fathered the concept, eventually became much more the interested observer offering a combination of imaginative and unconventional proposals whereas Harrison Brown became the organizer.

If one looks back on the various organizations of scientists which were generated between the period of 1945 and 1948 in response to the work of the Manhattan District as related in the comprehensive work by Alice Smith, *A Peril and a Hope*, one realizes that at that time Szilard was in one way or another deeply involved in the creation and guidance of most of them. He covered essentially all activities such as the organization and development of the Federation of Atomic Scientists, the Federation of American Scientists, the various movements at the national laboratories and universities, the action in Washington and the creation of the *Bulletin of Atomic Scientists*. In addition he aided in the collection of philanthropic funds and their distribution to the various operating groups.

There is little doubt that his basic motives were humanitarian and that his ultimate goal was to help in the formation of a universal government which would assure world peace. Some of those of us who worked with him, however, were occasionally put off by the fact that he was, initially at least, inclined to see no really fundamental difference between the truly democratic form of government and that of the Soviet Union. In fact, it seems safe to say that in spite of the freedoms which he enjoyed in an open democratic society, he distrusted the idea that the general public should have a strong voice in determining the course of events. On this score one might quote Eugene Wigner, one of Szilard's oldest friends, who stated in his memorial to Szilard in the Memoirs of the National Academy of Sciences: "It was a favorite saying of Szilard that one stupid person may be right as often as a bright one but two stupid people will be wrong much more often than two bright ones: They should not have as much to say about national politics as the latter. However his good will toward all including the stupid ones was always wholehearted and no one can accuse him of malice."

There are of course two great weaknesses to Szilard's approach to societal matters. First, it is very hard to get a representative group of even the most intelligent individuals to agree on relatively simple issues, let alone upon matters as complex as world government. Something in the nature of coercion by a selected few would be needed and we know where that can lead if appropriate good will and safeguards are inadequate. Moreover, the various meanderings of the United Nations Organization over the past decades do not give us complete confidence in the wisdom or steadfastness of any such international organization. Second, various groupings of people on our planet are highly diversified in essential ways and for one reason or another end up with different forms of government.

In the meantime, it appears that society must continue to carry on in what some may regard as a schizophrenic way by developing nuclear generated power where appropriate in order to provide essentially unlimited amounts of clean energy for the enrichment of our lives. At the same time, we must do our best to keep the destructive genie tightly sealed in the bottle. There appears to be no more rational approach to the complex situation which we face in the world in which we live today.

10
Science, Technology and the Environment

"The Professor is a person who thinks otherwise."

Old European Quip

Most populations which have been relatively deprived for an extended period and then experience an opportunity to move up the economic ladder, place first priority, in their own scale of values, on taking advantage of the new opportunities for such advancement. They almost invariably have secondary concern for environmental contamination as long as it is not an immediate and severe impediment to economic development. This trend seems to be almost international. The pea soup fogs and river contamination which characterized the earlier phases of industrial development in England are typical blights that populations will accept when new opportunities for wealth appear. Taiwan emerged from World War II as a relatively poor, primarily agricultural land but has since become a significant industrial center. The people on the island were prepared to ignore environmental problems and inconvenient congestion of various kinds while building up the industrial base and associated facilities, such as the transportation system and the institutions of higher education needed to benefit their children. The turning point has finally set in, however, and the government is establishing corrective measures.

The issue is probably not by any means restricted to modern times. The well disciplined and orderly rural and forest areas that one finds in most countries in northern Europe probably emerged as the standard only after the initial struggles to establish a viable agricultural economy had been achieved and many important lessons had been learned through painful experience.

Europe was also strongly influenced by the effects of military needs in the Middle ages when every community of any size was an armed fortress—a *Burg*.

This required detailed town planning and gave great power to architect-planners who still possess much consolidated influence. Such influence was essentially absent in the United States in any general way until very recent times.

1. The Start of the Environmental Movement

Since 1945 the United States has become acutely aware of environmental degradation after many decades of neglect. Perhaps the issue was brought to special public attention most vividly by Rachel Carson's book, *The Silent Spring*, published in 1961. However, clear recognition of the issue and significantly effective corrective measures had actually begun considerably earlier in many portions of the country. For example, western Pennsylvania, with Pittsburgh as its central urban center, became active almost immediately after World War II and took heroic measures to overcome many decades of neglect. The area had become the center of much heavy industry in the 19th century. This included establishment of facilities for the production of various forms of steel and other metal products as well as factories for the construction of heavy electrical equipment. Much wealth had been generated in the process. Some of the problems encountered in rectifying environmental degradation, but by no means all, were almost irreversible. For example, areas which had been subject to strip mining for coal and had been left barren without any reclamation turned out to be very difficult to rehabilitate in the short term. Most of the severe urban pollution was, however, corrected in fairly direct ways by going back to the sources of pollution and making significant alterations. New fuels for general use were introduced. Locomotives were converted to diesel engines or the system was completely electrified. Older contamination was cleaned up wherever feasible. This was accomplished with the joint cooperation of the city, state officials and industry. Concerned academic groups not only participated in the work in a constructive way but in some cases provided leadership in technical planning.

Much the same types of steps were taken in the area around London, England, in the same period. The pea soup fogs were eliminated and a start was made on clearing the Thames.

To appreciate the nature of the accomplishment it must be realized that the public at large in the greater Pittsburgh area had long associated pollution with prosperity in accordance with the belief that money and pollution go hand in hand. Clearly some economic price was paid for the changes through dislocation but all involved accepted them for the net benefits.

What occurred in western Pennsylvania was reflected by activity at more or less the same time in many other parts of the country. The Los Angeles basin with its special atmospheric conditions became conscious of what is termed "smog" and began to take action to control it. The same was true in the peninsula south of San Francisco and in Denver, Colorado so that the older industrial areas in the country were by no means alone in the movement.

The initial public action to rescue the environment in the United States was stimulated by two factors. First, it was finally appreciated that the country no longer had a significant geographical frontier bordering virgin lands into which individuals could migrate, leaving behind the debris, disorder and other forms of pollution generated by careless exploitation of previously opened new lands. Second, the population of the country was growing rapidly both through natural births and immigration so that there was reason to expect ever higher levels of pollution. In brief, the country was coming of age along with the realization that it had both a land and a heritage to preserve. A similar level of consciousness eventually emerges in most newly industrialized countries.

In this same period, essentially the 1950's, the public in the United States and elsewhere became aware of other factors that might affect health such as the heavy use of tobacco or diets too rich in fats.

2. The "Standard" Approach

One might have supposed that the examples of success achieved in Pittsburgh and many other communities wherein the constituents joined constructively with scientists and engineers to apply effective technical solutions to the control of pollution while encouraging economic growth might have led to unanimous agreement on the establishment of what could be called a national norm or "standard" for approaching such problems. In effect, one might have imagined an expansion of the authority of organizations such as the Occupational Safety and Health Agency (OSHA), or the creation of new ones, to cover ever-wider responsibilities beyond the industrial workplace but with a keen eye on the continuity of economic development. In fact this pattern of approach, which might be called the traditional one, is still probably favored by most individuals in the country. According to it the costs of achieving environmental resuscitation and control should, in the long run, be closely coupled to strengthening of economic productivity. Future actions would be extensions of the past with the great difference that environmental controls based on well-documented scientific and engineering research would be built into the ongoing system. Whatever economic or other costs might be incurred would clearly be balanced by well-defined gains in environment, health and aesthetics.

3. The Opposition

Actually, a very sharp division of opinion developed on this issue beginning in the 1960's, essentially twenty years after the country had first awakened in a serious way to environmental issues and had begun to take effective action. The 1960's proved to be a very turbulent decade in a number of other industrialized countries as well as in the United States with new groups arising to challenge what had previously appeared to be well-respected ideas.

The principal opposition group which has focused on environmental issues in the United States takes the view that the national style of living is basically wrong, being inherently profligate and wasteful at best. This is presumably compounded by the fact that advancing technology inevitably brings with it new and even greater threats to the planet in general and more specifically to mankind. Not least among the growing hazards, as they see them, is the ability of chemists to produce large quantities of new materials which pose great potential threats both to the environment and human health regardless of what benefits they may offer. Adequate cooperation of industrial organizations that are potential sources of pollution cannot be expected on any reasonable basis. According to this viewpoint, they will do their best to avoid the costs of repairing past damage or to treat ongoing production of wastes adequately. In other words, at the core, those supporting this point of view, which is closely similar to that of many members of the so-called "green" political parties of Europe, have no confidence in either the willingness or the ability of what is usually termed the establishment to control the situation.

As is normally the case in human affairs, the cohort of individuals supporting the opposition point of view is a mixed one. In the United States, as elsewhere, it has brought together, among others, the support of individuals such as dedicated nature lovers, those who dislike a consumer oriented society, those who would prefer far more central planning and control as well as those who find it a personally rewarding and perhaps even a lucrative cause with which to be engaged. To some of those involved in the opposition, the cause has taken on the aspects of a holy war.

4. Nuclear Power and Global Warming

At the start, the opponents focused on what might be called everyday environmental matters comparable to those that brought action to Pittsburgh and other communities about twenty years earlier. However, two other issues eventually became central to their mission, namely nuclear power and global warming. Members of the group support the view that in spite of its many advantages, including the fact that it offers an almost unlimited supply of energy, nuclear power is much too dangerous to be considered as a reliable or healthful source. With the use of skillful litigation that has raised the cost of constructing nuclear reactors by a factor of the order of ten, the group has succeeded in making it unprofitable for the electric power industry to expand its use of nuclear power even though about twenty percent of the electric power in the country is still produced in that way at the present time as a result of earlier installations. Only federal action, strongly supported, or even initiated by the public, could alter the situation.

While petroleum and natural gas are still available at reasonable cost and the country has an abundant supply of coal, closing off the nuclear option indefinitely promises to create serious problems in the 21st century. Both the cost and the availability of petroleum and natural gas will inevitably pose problems in a few decades at most. If the use of coal is eventually subject to some of the severe environmental regulations which members of the opposition group insist upon, this plentiful source of energy will not provide anything in the nature of a panacea. No alternative source of energy which could fill the gap created by the absence of nuclear and fossil fuels has yet emerged. While reference is often made to the use of solar energy, wind power or to controlled fusion based on the exploitation of the potential energy stored in deuterium, an isotope of hydrogen, these choices remain only distant possibilities at present and would probably generate problems of their own if they were to arrive on the scene in a broadly practical way.

5. Global Warming

The issues surrounding global warming are more complex and more difficult to assess in what might be called everyday practical terms. The atmosphere of the earth allows a large fraction of the light which reaches our planet from the sun to penetrate to the surface of the earth where it is transformed into heat. The atmosphere is so constituted that it impedes the escape of thermal radiation from the earth to space, acting somewhat in the nature of an insulating blanket. This is the so-called greenhouse effect. The natural greenhouse warming resulting from this effect corresponded to an increase in the surface temperature of the earth of about 30° C (54° F) prior to the industrial revolution. Various human activities such as the combustion of coal and petroleum products, which produces carbon dioxide, and the release of methane from various sources, add to the insulating effect of the atmosphere and to the natural greenhouse warming. If carried sufficiently far, the additional component could alter the climate of the earth significantly.

The issue is a very controversial one since we are not in a position at present to predict exactly how great the magnitude of the added effect would be, say at the middle or end of the next century, if the industrialized societies continue present practices. The most that can be done at present is to estimate the possible range of the effect. To add to the complications, the man-induced effect would be superimposed on significant natural variations derived from sources outside human control. A number of scientists have developed mathematical models of the processes involved in attempts to estimate the variations in climate from the various sources. However, these are not yet sufficiently precise (and may never be because of inherent complications) to give clear cut answers to these important questions. Apparently the best that can be said at present is that the additional rise in temperature would probably lie between essentially zero and about 2.5° C (4.5° F) by the end of the 21st century. One can say even less about the magnitude of the natural fluctuations over this period of time.

It is notable that we still do not understand the factors which caused great variations in climate in recent centuries, prior to the surge of industrialization. The climate was relatively warm nine hundred years ago when the Vikings roamed the northern seas in open boats. It became very cold in the 17th century when the canals in Holland froze in winter. The warming trend since can hardly be ascribed mainly to industrial emissions. Clearly there are significant natural causes for climatic drifts over periods of a century or so that are not yet understood.

Included among the many unknowns are questions regarding possible changes in ocean level, which rose approximately 100 m (330 ft) in the past twelve thousand years with the recession of the ice age. Also there is the question of whether or not a man-induced contribution to the greenhouse effect could forestall a future ice age.

In spite of the uncertainties in the theoretical estimates, there is a strong tendency within the opposition group to take the view that a significant level of man-induced global warming definitely will occur in the next century and that its effects will probably be widely harmful. Thus this group advocates that strong remedial action be taken as soon as possible on an international scale. Along with this point of view, the group also advocates that the United States should be among the leaders in cutting back on activities which produce greenhouse gases, most notably carbon dioxide.

6. International Cooperation

It is worth emphasizing that a very high degree of international cooperation on controls would be required if the movement to limit the man-induced emission of greenhouse gasses were to be effective, unlike the situation with respect to local pollution in which each nation or local region has relative freedom to decide the appropriate level of cleanup it desires as long as it does not affect its neighbors. In other words, to achieve success in controlling the man-induced component of the greenhouse effect, an international concordat of a remarkable kind would be needed. Moreover, it would have to be maintained essentially indefinitely without the substantial practical proof which is necessary. It will be interesting to see to what extent the international community will be willing to agree to such a plan, particularly when the costs, both direct and intangible, are appreciated.

One might have supposed that the growth of concerns with respect to global warming would have induced the opposition group to support nuclear power since it is not associated with the production of greenhouse gases. Any such change has been sporadic and on the part of individuals, but not sufficient to induce a truly significant shift in the attitude of the group as a whole. Some of those favoring nuclear power do so with the reservation that it should await the development of "new safe reactors" which has many of the aspects of seeking the end of the rainbow. It is not only that safety is a relative matter—both the French and the Canadians have a good record—but increasing safety in engineering systems requires conscientious working experience as well as theoretical groundwork.

7. Biological Concerns

It may be mentioned that one opposition group has focused on biological issues, including such matters as the use of pesticides in agriculture and the development of various species of living systems using laboratory-altered forms of genetic tissue (recombinant DNA research). While this group has achieved considerable success in its endeavors to exercise control over agricultural procedures and the use of genetically altered species, it is not yet an integral component of the mainstream protest group which focuses on issues emerging from the physical sciences. It does, however, enjoy considerable publicity.

8. Weak Electric and Magnetic Fields

The number of environmental issues about which one may be concerned seems endless. One of the very marginal ones, which has received a great deal of attention from a group of activists and much publicity by issue-oriented advocates in the media, relates to possible effects of the weak electric and magnetic fields associated with electric wiring in buildings, including homes, transformer vaults and high tension power lines. It is claimed that such fields may possibly induce cancer. It is suggested that children are at special risk for leukemia from this cause although there seems to have been no significant increase in the relative frequency of such leukemia over a period in which the national consumption of electric power has increased by a factor of ten. The effect, if any, lies fairly deeply buried in epidemiological statistics and may well be beyond any direct demonstration of either a positive or negative kind. In any case, the overall hazard would at worst seem to be very small when compared with the everyday hazards of life including those associated with automobile accidents. Nevertheless the matter can become a very contentious one when the siting of any office or school building is under discussion.

Thus far, the issue has not been made part of the package of concerns put forward by those who focus on nuclear power, global warming or gene splicing but the intensity of activity of those who are involved is no less.

It is perhaps worth mentioning that individuals residing in the developed countries have lived with electric wiring for at least a century during which there has been a most remarkable increase in health and longevity. Moreover, the original single-stranded form of electric wiring found in the older homes probably had much stronger fields than those derived from the present double-stranded system of wiring.

It appears at times as if the well-spring of superstitions which fed so many fantasies in the pre-scientific days has found new channels in which to flow in modern times.

9. Wealth of Support

It must not be supposed that what has here been termed the opposition movement is impoverished and without substantial economic or political power. It receives substantial funds from both public and private sources, including well-endowed private foundations and industry. Some of the contributions from industry clearly indicate general sympathy with the environmental movement even if there is not agreement on all details. Some, however, are apparently incidental to out of court settlements of litigation brought on by the opposition under the surveillance of sympathetic judges. Since both the print and electronic media enjoy publicizing controversy, the opposition groups have little difficulty in gaining publicity. This means, in turn, that they have little difficulty in obtaining the attention of the politicians.

One important measure of the success of the opposition has been the creation of an Environmental Protection Agency which is able to make recommendations for action as much upon what might be termed perceptions of possible hazard as upon demonstrable hazards based upon experiments or soundly documented epidemiological research. Presumptions can be given appreciable weight alongside factual observations in determining policy.

Since the overall costs of deviating significantly from what has been here termed the "standard" method of dealing with environmental issues, as advocated by the opposition, could undoubtedly be very high in intangible as well as tangible ones, the ultimate outcome of these debates will be notable to say the least. At stake are matters concerning the level of economic activity and the opportunities for the less privileged to climb the economic ladder.

What is important to emphasize here is that the opposition group, perhaps without explicitly realizing it, seems to be challenging the underlying spirit of the scientific revolution and its companion science-based technology to which the rise of science gave birth. Since the good and evil aspects of that technology are, in general, reverse faces of the same coin, the opposition group, if successful in all of its desires, might be led to place stringent controls upon scientific research at any level. That would usher in an entirely new era in the history of modern civilization.

11
Fraud, Piracy and Priority in Science

1. Fraud

In recent years there has been much discussion by the media, The United States Congress and in the general scientific journals of what is termed "fraud in science." We use quotation marks here because the issue can be a complex one in some cases. There are two general categories of what is termed fraud.

Monetary Fraud

The first category may be disposed of briefly. It relates to cases in which money given to support scientific research either directly or indirectly (so-called overhead) is used for quite different purposes. If the money has been provided by a governmental institution, there usually are laws governing the treatment of such fraud and, if discovered, it may become a matter for the judicial system. If the money originates from private sources, such as a private foundation, a corporation or an individual, it may also be subject to legal action depending upon the attitude of the donor. Beyond this, however, there may be a personal breech of faith which can have its own penalties. The best protection that the institution carrying on the research has against the perpetration of such fraud by the staff at any level is to have a good internal system of accounting and auditing and to establish firm rules regarding the dispersal of funds.

Scientific Fraud

What is usually termed scientific fraud by the media relates to attempts, whether successful or not, to publish in scientific journals or elsewhere material which the scientist carrying out the research has good reason to know is false. This is usually done to gain personal advantage or in support of some nonscientific doctrine. The most likely offenders are apt to be individuals who are at a relatively early stage in their careers and who hope to obtain a higher professional degree, an advance in rank or salary or, in some cases, merely to maintain a position which may be at hazard. Much more rarely, an individual with a well-established and well-earned position will engage in such fraud for reasons which are more difficult to judge but may be linked to the desire to gain some special recognition or, as mentioned above, to support a social or political cause in which he or she has a fervent belief.

If the perpetrator of scientific fraud desires to escape attention and achieve what might be regarded as a small, purely personal gain, it is important that the fraudulent result not deviate very much from what is regarded as conventional scientific opinion. It should not appear to be earthshaking or to represent what is termed a "breakthrough." Should the fraudulent result lie well outside the range of accepted views, it is almost certain to attract attention. It will be checked for accuracy elsewhere and the results questioned. In other words, the course of good science has a self-correcting quality if really major issues are involved. A fraudulent publication is most apt to get by if the material it contains is not of major significance and seems to fit in with well-established principles.

Protection

The greatest source of protection against such fraud is a well-developed system of peer review, particularly of the work of junior staff, within the supporting institution if that is feasible. Much "petty" scientific fraud is the result of pressure arising from the "publish or perish" syndrome and the fact that senior investigators may have more junior staff and students in their laboratories than they can supervise adequately on a personal basis. At this level, however, the potential that fraud will appear is more of an undesirable nuisance than a fundamental threat to the advance of science as long as the general review system remains open and free.

Occasionally a distinguished senior investigator will unwittingly become a partner to the generation of a fraudulent piece of work of a radical kind. In a not atypical case, the young investigator may emerge with some very preliminary results that appear to suggest the possibility of an interesting and perhaps even brilliant new finding. The senior investigator states: "If what you seem to find is true, I would suggest the following experiment." The young investigator, hoping to please the mentor, carries out the proposed experiment. Although the results actually are negative, the individual finds it convenient to claim fraudulently that the outcome was essentially as the senior investigator suggested might be the case. At this point the fateful die is cast unless the senior investigator participates in the work from that point on in a much more direct way. Otherwise, both individuals are headed toward disaster because the experiments are certain to be repeated elsewhere if the fraudulent claims are highly conspicuous and involve a suggestion of a revolutionary new viewpoint.

Swindlers

There are a number of instances in which senior investigators have been involved directly in work which becomes discredited. Some of these cases involve individuals who have no significant scientific credentials but are swindlers at heart, hoping to find gullible victims and achieve financial gain through intentional fraud. When seized upon by the popular media some frauds of this type can have a remarkably long lifetime.

Lysenko, Stark, Eddington, Newton, Kepler, Et Al.

Much more interesting and for our purposes more relevant cases are those in which the senior investigator has earned a solid reputation in one field and then decides to express very strong opinions with respect to one or more other fields. In many such cases it is not evident that the term "fraud" is actually appropriate since the individuals may be personally convinced that their results are correct. The Lysenko case described in Chapter 6 is an excellent example. In this instance, an agronomist who seemed reasonably capable in conventional terms attempted to discredit the validity of well-established principles of genetics and succeeded, with the help of the government, to gain national control of the field of plant breeding. A few previously distinguished German scientists, such as Johannes Stark, became so deeply involved in the racial theories of the National Socialist government that they expressed views in opposition to well-established scientific principles. Some of these individuals gained considerable power under Hitler although none ever achieved the dominance Lysenko did in the Soviet Union.

Another example relates to the brilliant astrophysicist, Arthur Eddington, who in his prime not only made major contributions to stellar theory but was a pioneer in advancing the general theory of relativity soon after Einstein's primary papers in the field became available. Apparently he realized when in his fifties that he would never have what might be called a complete understanding of the laws governing matter in the universe unless he devised them personally. As a result, he published in the 1930's and 1940's a most remarkable series of papers which turned out to have little to do with the real laws governing the universe. The issue here clearly is not one of attempted fraud in the true sense since there is little doubt that Eddington was convinced of the validity of his work at this stage. The case does, however, illustrate the degree to which a brilliant mind can be carried away with its own convictions.

Activities such as those in which Eddington became engaged near the end of his career are not uncommon for such brilliant individuals. Isaac Newton, for example, decided to use the special knowledge of astronomical theory that he had accumulated during his lifetime to fix the dates of major events in ancient history going back to the time of Noah and the Great Flood. His convictions concerning this work were very strong. However, it has not earned a permanent place in the field of ancient history.

There is also reason to believe that Kepler, having made the great discovery that the planets move in elliptical orbits rather than in epicycles, may have altered some of the experimental data available to him in order to fit the theory more closely than the experimental error associated with the measurements justified. The statistical treatment of experimental errors had not yet been developed in his day.

Perhaps the main conclusion that may be drawn from this is that if we exclude the problem of true monetary fraud of the type described in the first paragraphs, which may be dealt with through normal legal channels, what is called "scientific fraud" is not a matter of major significance. Unless it is linked to what might be termed trivial work which adds little stature to the reputation of the perpetrator, it will be discovered and discredited through the natural cleansing action of the processes involved in good science. No scientist of any age can hope to advance his or her cause in a truly significant way by engaging in fraud. This principle is widely understood in the scientific community. Once this has been said, however, it remains true that the community has a deep responsibility for maintaining systems of checks and balances which reduce occasional scientific fraud to the minimum possible.

2. Piracy

Alongside fraud as described in the previous sections, is the issue of piracy. It merits a comment here since it frequently attracts public attention at the same level as fraud and, when it occurs, is more often apt to involve fairly well-established investigators.

In a not unusual situation, a scientist who has made an interesting discovery submits a paper to a well-established journal describing the results of the research, or uses an account of the discovery prior to publication in a request for financial support to a government agency in order to extend the work. This is a very common situation in United States at the present time. An individual reviewing the material, either for the journal or a federal agency, hastens to write a brief note under his or her own name announcing the discovery and sends it to a journal that publishes such material rapidly and with a limited amount of review, thereby obtaining prior credit for the work. The pirate may or may not take the time to check the validity of the original work depending upon his or her confidence in the merits of the original research.

If the investigator responsible for the original work is a young, relatively unknown scientist, the consequences of such piracy can represent a significant setback, although his or her dilemma will become known to friendly senior colleagues. In any event, the news of the piracy eventually circulates through the scientific community and the pirate usually pays a price for the theft. One of the most unfortunate aspects of the potential existence of such piracy is that it may force an undesirable level of secrecy on investigators who would otherwise be much more open in discussing their work in the formative stage. It can also lead to hasty fragmented publication of results which could benefit from more thorough treatment.

Piracy is not very common since the vast majority of scientists respect the work of others. It is, however, an annoyance when it does occur. It is an open question whether the pirate gains as much by assuming false credit as he or she inevitably loses in reputation. The course of science as a whole is little affected by the existence of piracy but scientific life becomes somewhat less congenial because of it.

3. Priorities

It is a rare individual who remains unmoved if another assumes or is given credit for significant creative work of the former. Scientists, who are to a large degree motivated by the desire to seek new understanding of the natural world, tend to be especially sensitive to the issue of priority of discovery. Personal recognition from the professional community is a significant part of the reward they hope to obtain through their work. As a result, issues of priority loom relatively large among scientists and occasionally lead to spirited disputes. While one might suppose that such altercations are limited to individuals who are not highly creative and as a result must treasure every significant accomplishment, experience shows that this is not necessarily the case. Many justly famous scientists have proved to be as quarrelsome over issues of priority as lesser lights. Much depends upon the basic temperament of the individual. One could fill many pages with anecdotal accounts of such controversies involving scientists both great and small.

The most celebrated case of a priority quarrel, somewhat one-sided, is provided by Newton's ire on discovering the credit given to Leibniz for the latter's invention of the infinitesimal calculus. While there is no question that Newton definitely had prior claim, his tendency to keep his work secret made it inevitable that he would encounter a problem. The notion of infinitesimals was in the air and he could not expect to hold his secrets long. Very few scientific discoveries made from Newton's time onward could expect to remain hidden very long. Among the possible notable exceptions are Carnot's discovery of the entropy principle, Maxwell's formulation of the equations of the electromagnetic theory and Einstein's general theory of relativity. In another basic area, it took a full generation for Gregor Mendel's law of genetics to be rediscovered.

One French scientist remarked in humor that until recently it was rare to see a scientific paper originating in his country which had more than a single name on it because of the dispute which would arise among colleagues in the profession concerning *who* among the participants had *the* idea in the case of multi-authored work. In reality this comment can be applied to a degree across national boundaries without much modification.

While priority disputes fill the need of gossip channels very well and reveal the obviously human side of members of the scientific community, one may well ask if they are in any way a significant impediment to the advance of scientific research. The answer is definitely negative. Such disputes occur after the results have been achieved and represent no more than a side show in the main theater of science. At worst they may serve as a distracting element to those immediately involved in the dispute, and may disrupt previously warm friendships.

12
Big Science—Small Science

As we saw in Chapter 9, nuclear physics started its life in Montreal nearly a century ago as a form of what is termed bench science in which scientific ingenuity, a modest budget, access to supplies and a good instrument shop are the primary requirements for pursuing research. In the meantime it has revealed a vast subatomic world of fundamental particles of matter, along with knowledge that has contributed both to a deeper understanding of the structure of the atom and to technology. Unfortunately the scale of equipment needed to pursue the frontier opened by this research has grown essentially geometrically with each passing decade and has reached a stage at which each further advance requires facilities involving budgets which must be reviewed at the highest levels of the governments involved in supporting the research.

1. Reasons for Rising Costs

The reason for the rising cost is not difficult to understand in general terms. The primary charged particle probes used for exploration possess a characteristic wavelength that varies in an inverse manner with their speed. Since the degree of resolution of detail of the subatomic world is limited by the wavelength associated with the probe, much as the resolving power of an optical microscope is limited by the wavelength of light employed, advancing the field has required using particles of ever higher energies. This, in turn, requires increasingly larger and more expensive electromagnetic accelerators. In the very early stages of the research, Rutherford and his colleagues could use particles produced naturally by radioactive materials. Present-day accelerators produce particles having far higher energies. The so-called superconducting supercollider being designed for a site in Texas, which can be regarded as a giant microscope of a kind, will cost in the neighborhood of ten billion dollars and will need an annual operating budget of the order of several hundred million dollars. The use of the machine will be shared by many groups under the guidance of what is termed a user committee composed of scientists.

2. Other Expensive Fields

High energy particle physics is not the only area of science that requires a significant amount of money for large new equipment if the frontier of understanding is to advance. Forefront telescopes become larger and more elaborate as the astronomers look farther into the universe and seek increasingly detailed information. Moreover, it is highly desirable that new instruments be situated in fairly remote places to gain special advantages, including isolation from urban effects such as smog and city lighting and relative freedom from atmospheric turbulence. Such isolation involves greater supporting and operating costs. The leading ground-based telescopes cost tens of millions of dollars. The orbiting Hubble telescope cost about one billion dollars, exclusive of the cost of maintaining the operations of the launching site.

In many ways, the surface of the moon would provide an ideal site for a major astronomical complex. From it one could, among other things, hope to study aspects of the planets of neighboring stars completely free of atmospheric turbulence. It would probably require human servicing and involve a cost on the scale of the Apollo program of the 1960's, along with an equally significant operating budget. Clearly such a base could be used for a variety of other scientific and technical purposes in parallel with the astronomical activity, but the added cost for astronomical work would still be very large.

On the biological side, the cost of the human genome project which is designed to map the basic pattern of the human genetic system, is estimated to be three billion dollars. It may be anticipated that this program will turn out to be the start of an increasingly elaborate and useful study of human physiology at the molecular level.

3. Continuing Importance of Bench Scale Research

In the meantime, bench scale research in both the physical and biological sciences has lost none of its importance or value, granting that the string and sealing wax era of investigation is probably gone forever at the most creative frontiers. As was mentioned in an earlier chapter, equipping even a modestly adequate research laboratory for an investigator and the associated group of junior colleagues at the bench level now requires several million dollars. Much highly productive research in fields such as chemistry, biology, condensed matter physics, optics and geology is carried on in laboratories funded at this relatively modest scale. Depending upon the field, a combination of efficiency and effectiveness make it expedient to use items such as centrifuges, protein sequencers, computers, electron microscopes, high field magnets, microwave generators, laser sources and the like, often obtained in significant part from commercial sources. This does not mean that access to multi-million dollar facilities such as neutron sources or equipment for producing extremely high intensity magnetic fields cannot occasionally be useful in such work. They, however, are usually employed on a shared basis in what might be termed the occasional user mode.

4. Conflict

Since national budgets for scientific research, while somewhat flexible, do have limits and since private sources, while exceedingly useful, are inadequate to provide more than a relatively small fraction of the funds needed to support the most basic research, it is inevitable that friction develops between groups requiring very expensive specialized equipment and those engaging in smaller bench-scale research, particularly since the latter can point out how much further the money spent for large-scale science would go if devoted to their type of work. In addition, the academic scientists of this type emphasize that the vast majority of those who receive advanced degrees in science do their work with them rather than with those working in big science.

This conflict emerged in a significant way in the 1960's when big science gained prominence. The conflict clearly will be with the technically advanced societies indefinitely in the future. Much can be said on both sides of the controversy since both groups can boast of remarkable achievements and offer vistas of promising results in the future.

With regard to large scale expensive science, it would appear to be a reasonable part of the human quest for knowledge to obtain deeper understanding of the ultimate nature of the matter in the Cosmos, as well as of the factors in our physiological constitution that make us what we are, to the extent our means and capabilities permit. On the other hand, bench level research will remain at the core of the scientific endeavor indefinitely in the future because it provides the means for dealing with the essential details of so much of the biological and physical world. As was mentioned in Chapter 7, the field of biochemistry is still an open book. We have barely scratched the surface in the domain of neurophysiology. Fields such as chemistry and condensed matter physics continue to yield very rich dividends with no indication of significant limitations. Moreover even the grandest form of large-scale science has a significant component of bench science at its base, whether for the development of essential auxiliary equipment or in the analysis of data.

5. Mediation

How is this issue to be mediated? Since there is substantial merit on both sides, the resolution requires compromise while recognizing that no rigid formula will prove adequate. A pattern of approach that seems to work reasonably well, and actually is followed to a substantial degree by the leading granting agencies in the United States, is as follows. Its success, when used, depends upon close cooperation between working scientists and those who oversee the actual distribution of funds.

Each subfield of science should, on a more or less continuing basis while being guided by those who work in the field, establish its own system of relative priorities as best it can. In general, the pattern of priorities which emerges will, at least in its early stages, have more than a single column because of the inevitable diversity of activities in most subfields. However they may be derived, such priorities can then form the basis for establishing priorities in the major fields of science such as mathematics, astronomy, physics, chemistry, geology and the various branches of the life sciences, again with recognition of the essential diversity within each field.

It should not be assumed that the process of setting priorities in this way, with the use of professional scientists, is an easy one. A long chapter could be written on this subject. It took the biochemists almost a decade after the great discoveries that were made in the 1940's and 1950's to achieve appropriate recognition in the field of chemistry, although they were greatly aided by the fact that their research has so much relevance to medicine. They have since attained great prominence. In fact the basic support at the academic level of organic and analytic chemistry has probably dropped below the ideal level. It is notable that the physical chemists, who achieved substantial recognition in earlier decades of the century after something in the nature of a struggle for recognition, have found what might be termed refuge among the condensed matter physicists and the biochemists to a significant degree.

The support provided to the field of condensed matter physics appeared to have levelled off in the 1960's and 1970's after the elucidation of the type of superconductivity that is observed in a number of metals at temperatures in the range associated with liquid helium and liquid hydrogen. The discovery of high temperature superconductivity in the 1980's has, however, given the field new impetus.

Geophysics, which was once a field of small science, has developed aspects which lie in the realm of big science. The verification of the concept of continental drift has opened up the field for observations on the global scale.

The high energy physicists supported almost unanimously the view that priorities should be given to the development of new machines that produce bombarding particles having increasingly higher energies even at the expense of intensity, although intensity is always valued. Controversy developed at the professional level, however, regarding the design, location and management of a major new accelerator during the procedures which eventually led to the establishment of the Enrico Fermi National Laboratory in the state of Illinois. The European high energy physicists seemed to have had less controversy in connection with the evolution of the great laboratory (CERN) near Geneva, Switzerland, perhaps because of constraints concerning siting.

An interest in cosmological science has dominated the fields of astronomy and astrophysics for a number of decades. While this is easy to understand in view of the remarkable discoveries being made in the field, it is quite possible that more attention should be given to solar physics because of its relevance to matters close to our own planet. It would be good, for example, to have more quantitative information on the variations in output of solar radiation both annually and over decades.

The support of aspects of science which have an immediate bearing on the fate of the environment can generate great emotion among professional groups. Scientific objectivity can suffer to a degree in the process. Both the politicians and the media have learned that they can easily generate heated debates among appropriately selected groups of scientists by opening such subjects to discussion.

Once the active scientists have given their opinions, it is necessary to rely on the wisdom and experience of the major science administrators and those who advise them in seeking a balance among contending groups. It is inevitable that contingencies and irrelevancies will play a role to some degree at this stage as they do in most human affairs. Indeed, this is almost guaranteed by the fact that the support of science is now so intimately intertwined with national political and economic policies.

A key to the success of any system, such as that described here, lies in its ability to preserve flexibility so that there is always an opportunity within a reasonable time scale to readjust matters if and when an unanticipated opportunity or set-back arises.

One must retain the hope that the highest levels of society will continue to encourage the advancement of science while leaving the major decisions concerning the balance of priorities in the hands of experienced scientists. To paraphrase a statement often made with respect to the democratic system in the field of politics, the process of decision-making in science described above, is not necessarily a good one but it is the best available to the degree that it relies upon the opinions of working scientists.

6. An Alternate View

It has been suggested at various times that the support provided to a given field of science should be based primarily on an assessment of the importance of that field to other fields of science. The proposal supported here, which actually is in fairly general use at present, contradicts such a policy at least at the primary level of the decision process. Each area of science has its own internal consistencies and challenges which are best understood by those working in the field. The process of establishing priorities must start with the advice of those investigators. The issue of cross relevance to other fields should, if it enters at all, occur at a much later and higher level.

It is common experience that what may at one time appear to be a highly esoteric field of research which has relevance only to itself will, at a later time, prove to be exceedingly important for another field of science—in some cases for a field of technology. What were once regarded as among the most abstract areas of mathematics have eventually provided everyday tools in the fields of science and engineering.

We shall end with an anecdote relevant to this. When Professor Albert Claude first presented to a gathering, at what is now Rockefeller University, his plans for disrupting cells, separating the organelles with a centrifuge and studying the constituent parts both chemically and with the use of the newly invented electron microscope, a distinguished member of the audience rose and said: "I do not know what you will eventually get from all this but I am certain you will start out with a horrible mess." Modern molecular biology emerged from that unlikely mess.

Name Index

Subject Index

Leukemia, 125
Lick Observatory, 53
Lillienthal-Atcheson Report, 115
London, 120
Lunar Observatory, 73, 134

Mainland China, 74, 78
Malthusianism, 83
Man as a wild animal, 86
Manhattan District, 64, 115
Manichaeism, 27
Many body systems, 100, et seq.
Martyrdom, 94
Massachusetts Institute of Technology, 51
Maxwell's Demon, 91
McGill University, 43, 111
Mendelian genetics, 78
Metabolic stability, 104
Milk producing hormones, 79
Mineral exhaustion, 89
Mitochondrion, 99
Molecular beams, 103
Molecular biology, 106, 109, 138
Monotheism and science, 39
Morrill Act, 47, 54
Moslem science, current, 74
Muir, John, 48

National Academy of Engineering, 60
National Academy of Sciences, 47
National Advisory Committee on
    Aeronautics, 60
National Cancer Institute, 62
National Defense Research Committee, 63
National Institute, 45
National Institutes of Health, 62, 64
Nationalism, rise of, 30
National Physical Laboratory, 83
National Pride, 73
National Research Council, 60
National Science Foundation, 64, 68
National Socialism, 63, 129
Natural selection, 102
Naval Research Laboratory, 61
Near East Conflict, 95
Neoplatonism, 22
Netherlands, 31
Neurophysiology, 135
Newton, religious beliefs, 37

Nobel Prize, 56, 57, 61, 63, 111
Novum Organum, The, 35
Nuclear energy, 16, 79, 88, 117, 122, 124
Nuclear fear, 109
Nuclear fission, 63
Nuclear fusion, 123
Nuclear physics, 43, 133
Nuclear science, 63

Ocean level, 124
Office of Naval Research, 64
Office of Scientific Research and
    Development, 63
Oil, spreading on water, 44
Open society, 78
Organelles, 97, 138
Overpopulation, 85
Oxford University, 26

Padua, 34
Paris, University of, 26
Pea soup fogs, 120
Persian science, 24
Phoenicians, 11
Piracy in science, 130
Pittsburgh, 120
Portuguese, 10
Princeton University, 46, 50, 54
Printing, invention of, 30
Priorities in science, 131, 135
Prokaryotic cells, 98
Proteins, 98
Public Health Service, 45, 62
Public interest, 76
Public Works Administration, 63
Pugwash, 115
Pythagorean Theorem, Chinese proof, 38

Quantum mechanics, 4, 100, 103

Radioactivity, 109, 110
Radiochemistry, 104
Rainbow, 2, 29
Recombinant DNA, 79
Reductionism, 99
Reductionist analysis, 81
Relativity, 4
Renssalaer Polytechnic Institute, 53
Rockefeller Foundation, 61